Essentials of Supersonic Commercial Aircraft Conceptual Design

Aerospace Series

Essentials of Supersonic Commercial Aircraft Conceptual Design

Egbert Torenbeek
Delft University of Technology
Netherlands

This edition first published 2020
© 2020 Egbert Torenbeek. Published 2020 by John Wiley & Sons Ltd.

The right of Egbert Torenbeek to be identified as the author of this work has been asserted in accordance with law.

Registered Offices
John Wiley & Sons, Inc., 111 River Street, Hoboken, NJ 07030, USA
John Wiley & Sons Ltd, The Atrium, Southern Gate, Chichester, West Sussex, PO19 8SQ, UK

Editorial Office
The Atrium, Southern Gate, Chichester, West Sussex, PO19 8SQ, UK

For details of our global editorial offices, customer services, and more information about Wiley products visit us at www.wiley.com.

Wiley also publishes its books in a variety of electronic formats and by print-on-demand. Some content that appears in standard print versions of this book may not be available in other formats.

Library of Congress Cataloging-in-Publication Data
Names: Torenbeek, Egbert, author.
Title: Essentials of supersonic commercial aircraft conceptual design /
  Egbert Torenbeek, Delft University of Technology, Netherlands.
Description: First edition. | Hoboken, NJ : John Wiley & Sons, Inc., 2020.
  | Series: Aerospace series | Includes bibliographical references and
  index.
Identifiers: LCCN 2019053571 (print) | LCCN 2019053572 (ebook) | ISBN
  9781119667001 (hardback) | ISBN 9781119667032 (adobe pdf) | ISBN
  9781119667049 (epub)
Subjects: LCSH: Supersonic transport planes–Design and construction.
Classification: LCC TL685.7 .T67 2020 (print) | LCC TL685.7 (ebook) | DDC
  629.133/349–dc23
LC record available at https://lccn.loc.gov/2019053571
LC ebook record available at https://lccn.loc.gov/2019053572

Cover Design: Wiley
Cover Images: Supersonic plane © JimboMcKimbo/Shutterstock,
                The nose of the concord © Kenneth Summers/Shutterstock

Set in 9.5/12.5pt STIXTwoText by SPi Global, Chennai, India

Printed and bound by CPI Group (UK) Ltd, Croydon, CR0 4YY

10 9 8 7 6 5 4 3 2 1

This book is dedicated to my wife Nel Torenbeek-Volker with sincere thanks for her patience and stimulus during more than six years of hard work.

Contents

Foreword

Egbert Torenbeek's book on commercial supersonic aircraft design is timely, exactly 50 years after the first flight of the Concorde in 1969. It performed scheduled flights for 27 years and was able to operate with a profit because the aircraft's attractiveness was able to sustain a high ticket price. As such, the Concorde is the reference aircraft in Torenbeek's book, which starts with a chapter on Concorde's development and service. Torenbeek believes that "a new generation of supersonic passenger aircraft could have a commercial future a decade from now". From this the author takes his motivation. He writes for a potential engineering team producing a conceptual design for a supersonic airliner. For his wider readership Torenbeek digests the best of the available literature and puts it together in a concise form. He draws his own books and papers on aircraft design and quotes authors that were on the forefront of supersonic aerodynamics: L. Prandtl, J. Ackeret, M.M. Munk, T. Von Kármán, A. Busemann, D. Küchemann, R.T. Jones and J.D. Anderson Jr. Moreover, some knowledge from the ESDU Data Sheets is used. Although many books are available about supersonic aerodynamics and supersonic design, Torenbeek puts all this together and writes about supersonic commercial aircraft design. In Chapter 8 about aerodynamic efficiency of supersonic cruise vehicle configurations, the pros and cons of configurations are compared, in particular the aft tail, foreplane and tailles designs. Clearly, the book has an aerodynamic focus as the aircraft cruise speed is supersonic, but the aerodynamic aspects are always discussed from a design perspective. This is especially true for Chapter 3 about weight sensitivity and energy efficiency, where Torenbeek starts this item with the first law of aircraft design, which states that the sum of the payload fraction, the empty weight fraction and the fuel fraction is equal to one. This equation also shows that not every design problem will have a solution if technology parameters for lightweight design and/or fuel weight are suitable. In the case of the Concorde, the maximum payload is only 6%, its empty weight fraction is 44%, and the relative fuel mass fraction is 50%. This is not a favorable comparison with the relative 25% for subsonic short-range passenger aircraft and 10% for subsonic long-range aircraft.

In Chapter 4 Torenbeek writes: during the development of the Concorde, devoted proponents suggested that the fuel efficiency at supersonic speed is not very different from the fuel efficiency at subsonic speed, arguing that that the deterioration of the deterioration of L/D at supersonic speed caused by supersonic wave drag is compensated by the high Mach number. Here the proponents used the term ML/D as the factor determining the fuel efficiency. Torenbeek points out that this is not correct since the total effects should be determined by the Breguet range equation, specifying that the range is proportional to ML/D divided by the specific fuel consumption TSFC of the installed engines, which is considerably higher at supersonic speed compared to subsonic speed. The author celebrated this year his 80th birthday and this foreword would be incomplete without looking back at his achievements. Egbert Torenbeek studied at the Delft University and graduated with a degree in aeronautical engineering. In 1961 he took the Guided Missiles Course at the College of Aeronautics in Cranfield (UK) which was followed by his military service in the Dutch Air Force from 1962 to 1963. He supervised a teaching course in the TU Delft to start working under Hans Wittenberg, professor of aircraft design. Torenbeek supervised a teaching course and concluded that there existed no up-to-date handbook on aircraft design. So, he collected information that had been published up to 1970, when passenger airplanes such as the DC-8, the Boeing 707 and the Lockheed Tristar were already operational and Concorde had made its first flight. After about six years of work the book Synthesis of Subsonic Aircraft Design was published by the Delft University Press in 1982 and is presently distributed by Springer. After a sabbatical period of nine months in 1977 at Lockheed Georgia (USA), Torenbeek became full professor in 1980. In 1993 he had the leadership of the EXTRA 400 conceptual design, which was made with the engineering help of tests in the wind tunnel carried out at the TU Delft. The LBA Type Certificate was obtained in 1997. Torenbeek was the co-founder of the European Workshop on Aircraft Design Education (EWADE), which is held every two years and included one day for informal discussions where new ideas were discussed in a nice setting. The Journal Aircraft Design was started by Elsevier in 1998. Egbert Torenbeek and Jan Roskam acted as editors in chief. Torenbeek served two years as vice-rector and continued as professor emeritus. His early retirement was closely related to political discussions in the wake of Fokker's bankruptcy. In 2000 he received an honorary doctorate from the Moscow Aviation Institute, which he sent back in 2014 as an act of protest immediately after the MH17 disaster. The book Flight Physics (co-authored with H. Wittenberg) was published by Springer in in 2009. His book Advanced Aircraft Design was published in 2013 by Wiley and translated into the Chinese language. In 2013 the Aircraft Design Award from the American Institute of Aeronautics and Astronautics (AIAA) was given to Torenbeek and in 2016 he received the Ludwig Prandtl Ring from the German Society for Aeronautics and Astronautics, which is awarded for an outstanding contribution to the field of aerospace engineering. Torenbeek presently acts as Honorary Guest Editor for the Continuous Special Issue Aircraft Design of the journal Aerospace at MDPI.

What will the future bring for supersonic commercial transport? Several supersonic business jets are in the design stage, whereas several such projects have already been given up. It is difficult to get the economics right. Development costs to cope with technological challenges will be high and numbers produced in the end will be rather limited. Currently, the US law prohibits supersonic flight over land unless authorized by the FAA for purposes stated in the regulations. There are supersonic rule-making activities, but none of them would rescind the prohibition of supersonic flight over land. Environmental questions remain due to high fuel consumption in the stratosphere and the considerable take-off noise produced by Concorde will have to be considerably reduced, although the last chapter promises to have a possible solution for the conceptual design problem. First of all, it is important to understand the essential conceptual design concepts. This book by Egbert Torenbeek delivers this knowledge.

1 June 2019   *Prof. Dr.-Ing. Dieter Scholz*
                    MSME. Hamburg University of Applied Sciences Hamburg

Series Preface

The field of aerospace is wide ranging and covers a variety of products, disciplines, and domains, not merely in engineering but in many related supporting activities. These combine to enable the aerospace industry to produce exciting and technologically challenging products. A wealth of knowledge is contained by practitioners and professionals in the aerospace fields that will benefit other practitioners in the industry, and to those entering the industry from University.

The Aerospace Series aims to be a practical and topical series of books aimed at engineering professionals, operators, users, and allied professions such as commercial and legal executives in the aerospace industry. The range of topics is intended to be wide ranging covering design and development, manufacture, operation, and support of aircraft as well as topics such as infrastructure operations, and developments in research and technology. The intention is to provide a source of relevant information that will be of interest and benefit to all those people working in aerospace.

This book extends the author's previous excellent and informative treatises on concept design to focus on supersonic transport aircraft for commercial use. The heady days of supersonic aircraft designs from the UK, USA, and USSR are long gone with the demise of SST for a number of programme and operational reasons, largely related to operating and support costs. A surge in leisure and business travel together with savage competition to reduce ticket prices led to the emergence of very large aircraft and ETOPS which made long distance travel relatively comfortable and affordable. This, and an increase in e-commerce and environmental concerns, seemed to indicate that the days of supersonic business travel would never return. However, modern business and diplomacy still requires face to face discussions and rapid responses that can be made easier with supersonic travel, so there is a potential market, if not for mass travel then certainly for business users for whom time is valuable.

The author has taken a practical view of the possible re-emergence of supersonic transport by examining the history of the previous projects and the lessons to be learned from that era. He has developed this by considering contemporary circumstances that have an impact on concept design, such as modern design methods, modern materials, modern aircraft shapes, and environmental issues. This provides a fund of invaluable experience and material for concept designers to examine the engineering and operational aspects of supersonic transport. This knowledge will contribute to the study of the feasibility of supersonic commercial types that will be viable in today's economic and political climate. This is a valuable book for concept engineers, politicians, operators, academics, and forward thinkers.

November 2019 *Peter Belobaba, Jonathan Cooper and Allan Seabridge*

Preface

This book is intended to be used by members of a team producing an initial design concept of an airliner with the capability of making supersonic cruising flights. Since the demise of the Concorde more than half a century ago there are no designers left with the experience and knowledge required for developing a new initial design proposal. On the other hand, since Concorde's birth there has been a wealth of scientific publications on topics, such as the development of supersonic cruise vehicles, aerodynamics, propulsion, structural design, and flight physics, and in particular the analysis of the sonic boom. Moreover, there appears to be a considerable niche market for relatively small high-speed aircraft, in particular business jets.

The development of supersonic technology since the end of the twentieth century has primarily advanced in the field of transonic and supersonic aerodynamics. For example, many studies have been carried out in the field of configurations with oblique wings, promising improvements in flight efficiency of up to 20% as well as large gains in reducing the sonic boom, take-off noise, and low-speed performance improvements. From this point of view, a new generation of supersonic passenger aircraft could have a commercial future a decade from now.

Although the present generation of aircraft designers has enjoyed an introduction to the physics of supersonic flows during their academic education, not many of them have actually experienced activities associated with the design of a supersonic cruise vehicle. Fortunately, a wealth of high-quality information on applied supersonic aerodynamics is available in classical books such as the well known books of D. Küchemann, J.D. Anderson and D.L. Raymer. Together these texts provide a comprehensive introduction into the fundamentals, and analytical and computational treatment of high-speed flows.

Acknowledgements

The author is indebted to Professor Leo Veldhuis for his hospitality by offering me a room to continue writing this book during many years after my retirement. Moreover, many colleagues have assisted me in solving problems in the application of the LATEX program, in particular Dr. Roelof Vos, Dr. Maurice Hoogreef and Ir. Reno Elmendorp.

1

History of Supersonic Transport Aircraft Development

At the end of the 1950s military jet aircraft made routine flights at speeds faster than sound, and the first generation of long-range high-subsonic jet-powered airliners had only just been introduced into service, when it was realized that supersonic airliners could become a reality. The commercial potential for supersonic flight came under serious study in the four nations that fostered their development: France, UK, USA and USSR. Companies in the USA coupled experience obtained from the development of military vehicles during the 1950s (B-58 Hustler, B-70 Valkyrie) with successful jetliner programs in order to develop a supersonic transport (SST) designed to travel at up to three times the speed of sound in the stratosphere. Its funding required direct government sponsorship, with a series of competitions, selecting Boeing as the airframe manufacturer and General Electric as the engine manufacturer. Due to a variety of economic, environmental, and political issues, the development of the Boeing 2707 prototype was discontinued in 1971, nine years behind schedule and 20% above design weights. In 1962 an Anglo-French consortium consisting of the British Aircraft Corporation (BAC) and Sud Aviation started the development of the Concorde. Almost concurrently the Soviets revealed that they were developing a supersonic transport in a manner conventional to their style, with the government assigning the project to Tupolev. Both aircraft (Figure 1.1) were designed to fly at approximately twice the speed of sound (Mach 2). The TU-144 made its first flight in January 1969, was introduced into service in 1977 but suffered from excessive fuel consumption and severe operational difficulties. Since it was apparently unsafe and considered virtually useless, the first TU-144 was withdrawn in June 1978 after 55 scheduled flights. Commercial transport at supersonic speeds was a reality from January 1976, when Concorde entered successful commercial service for 27 years with British Airways and Air France. It is therefore stunning that many "experts" have considered the Concorde a great technical achievement but an economic disaster.

Essentials of Supersonic Commercial Aircraft Conceptual Design, First Edition. Egbert Torenbeek.
© 2020 Egbert Torenbeek. Published 2020 by John Wiley & Sons Ltd.

Tupolev Tu 144   BAe-Aerospatiale Concorde

Figure 1.1 The only supersonic commercial aircraft serving in commercial operations. Courtesy: Flight International.

1.1 Concorde's Development and Service

Early design studies in the 1950s by the UK industry aimed at a supersonic airliner designed for non-stop flights between London and New York. One concept was equipped with a slender body and very thin straight wings, not unlike the general arrangement of contemporary supersonic bombers. This configuration could not generate an acceptable aerodynamic quality, resulting in an aircraft carrying only fifteen passengers with a take-off weight of 136 metric tons. The large wave drag of its wing was the major obstacle for efficient flight and aerodynamic experts at the Royal Aircraft Establishment (RAE) soon realized that wave drag could be kept low by using a slender wing to keep the leading edge behind the Mach lines from the vertex.

In 1956 the RAE and aircraft manufacturers established the Supersonic Transport Aircraft Committee (STAC) with the intention of taking the lead in designing and producing SST. The STAC concluded that most operational advantages of supersonic long-range flying were secured if the vehicle cruised at a speed near 2000 km h^{-1} (Mach 2), which would enable the airline to fly two transatlantic round trips per day. Moreover, at this speed the kinetic heating of the structure would allow the use of advanced light alloys instead of steel or titanium required for Mach 3. In 1960 Bristol Aircraft was awarded a contract for designing a supersonic commercial transport (SCT) for 130 passengers, which was completed in 1961.

Around the same time the French air ministry requested a proposal from aircraft manufacturers for a medium-range SCT cruising at a Mach number between 2.0 and 2.2 with a capacity of 60–80 passengers. ONERA was selected for basic theoretical and experimental research and the resulting projects by Sud Aviation, Dassault and Nord Aviation were completed in 1961. The French officials concluded that the Sud design was the most promising. Despite the different payload and range requirements, the British and French teams evolved broadly the same aerodynamic design approach and it was realized that they should collaborate in a project that would benefit both industries, and the same applied to the participating British and French engine industries. After consultations with potential customers and governments it was decided that the Anglo-French supersonic transport would carry 130 passengers over the Paris–New York Atlantic range. The formal Anglo/French agreement for development and manufacture with a production line in both countries was signed in November 1962 and prototype construction began in 1965.

The aircraft, baptized "Concorde" produced by BAC and Aerospatiale, made its first flights in early 1969. A total of twenty aircraft were constructed, including two prototypes and two pre-production models. Fourteen of the sixteen series-produced aircraft served mainly on North Atlantic routes, split between British Airways and Air France. They carried their passengers cruising at speeds up to Mach 2 at 18,000 m altitude and thereby saved four of the typical seven hours trip time required by high-subsonic jetliners. However, Concorde was developed just prior to the establishment of FAR 36 noise regulations and – with its afterburners operating during the take-off – the aircraft required a noise rule waiver to allow its operation out of American airports. Moreover, the establishment of FAR 91 rules in 1973 prohibited sonic booms over inhabited areas, making flight at Mach 2 over these areas impossible. It was not until 1980 that Concorde reached the point where it could carry a full load of hundred passengers year-round on the North Atlantic routes.

The Concorde and Boeing SST programs were conceived at a time when fuel prices were coming down. However, supersonic cruise requires more energy per unit of payload and range, and both designs were known to be sensitive to the availability of fuel. Due to the oil crises in the 1970s and the subsequent increase in fuel price as well as the increasing concerns about the effects of supersonic flight on the environment, the interest in supersonic civil aviation decreased and Concorde remained the only SCT in regular airline use during the twentieth century. Scheduled flights were principally London–New York and Paris–New York and they attracted mostly high utilization. During the 27 years of their operational life a fleet of only twelve flying Concordes accumulated some 350,000 hours, most of the time flying at supersonic speed – more than all of the world's military aircraft together – and with high reliability. During the years of Concorde's operational life, it was generally concluded by British Airways and Air France that, despite its high maintenance costs, the technology generally satisfied or exceeded the expectations at the start of the project.

In August 2000 a piece of titanium left on Charles de Gaulle Airport's runway caused Concorde's landing gear tire to explode, damaging its wing fuel tank structure and setting an engine on fire. After lifting off, the plane could not climb out, became uncontrollable, and crashed. Although British Airways and Air France considered the Concorde to be profitable up until the accident, they concluded in 2003 that continuation of its services was no longer commercially justified. In particular, the high fuel costs per seat-kilometer, the maintenance costs of seven times those of a Boeing 747, and the modification costs expected in that year were behind the decision to phase out its operations. Economically, Concorde did not fit into the structure of the air traffic system due to its high operational costs, and the high research and development costs could not be negotiated by the small number of aircraft produced and sold.

In spite of its high cruise speed reducing the time to travel drastically, and the fact that it provided a safe and reliable Atlantic service from 1976, Concorde is sometimes portrayed as a folly and a failure, but this ignores the fact that the USA once viewed it as a threat to its aerospace leadership. The Concorde was a technological and systems integration marvel in its time – an achievement that since its emergence has never been surpassed. Its development, production, and service have enriched the knowledge of European technological cooperation. Apart from the excellent flying qualities demonstrated during its service, the Anglo-French supersonic transport was the first international aerospace program that reshaped industrial and political thinking and it paved the way for most European collaborative programs. Its legacy is today's European aerospace industry Airbus, established in 1970, and the European certification authority EASA.

1.2   SST Development Program

The efforts in the US to develop a supersonic airliner were preceded by a comprehensive program of supersonic military aircraft development. From the early 1950s the Air Force operated the Convair B-58 Hustler Mach 2 bomber and the North American XB-70 Valkyrie bomber/reconnaissance aircraft (Figure 1.2) was conceived during the late 1950s. The design specifications of the B-70 were influenced by the opinion of military authorities that its high cruise speed should be approximately Mach 3 at 21,000 m altitude, since it was anticipated that the additional research for achieving the same flying qualities as for Mach 2 would be modest. However, aluminum alloys could not be used due to the strong kinetic heating effects of flying at Mach 3 and hence alternative structural materials such as stainless steel and titanium had to be incorporated. Test flying of the XB-70 demonstrated that it had excellent aerodynamic qualities in supersonic flight as well as acceptable low-speed characteristics. Although the B-70 program was canceled for strategic reasons after three prototypes had been built and tested, arguments behind the development of a Mach 3 airliner were dominated by the experience gained during research of the XB-70, and Boeing initiated a design study of an SST, which in 1952 resulted in the Boeing 2707-300.

NACA's supersonic commercial air transport (SCAT) research program was initiated in 1957. Initially there was no government support for a CST development program. However, as soon as the European plans for producing the Concorde

Figure 1.2 The North American XB-70 Valkyrie strategic bomber/reconnaissance aircraft (first flight made in 1964) [4].

appeared to be taken seriously by the airlines, Pan Am wished to be "the first airline to go supersonic" and placed options to buy six aircraft. As one result of this challenge to the "free enterprise American industry", the development of an SST prototype was addressed by President Kennedy in 1963 as a national objective. The FAA was designated to conduct a design competition between Boeing, Douglas, and Lockheed for a full-scale pre-production SST prototype program. Financial support by the USA government for the project was assured for a program whereby 90% of the funding came from the government and the remaining 10% from the industry. The government's investment would eventually be returned from the aircraft's proceeds of sale.

The American SST projects of the late 1960s and early 1970s aimed at carrying more than twice as many passengers as the Anglo-French Concorde over considerably longer distances. Concorde's competitors initially chose an aggressive Mach 3 cruise regime for the US transport market, similar to the military supersonic cruising vehicles. NASA directed a competition between proposals generated by Boeing, Lockheed, and North American. Featuring a variable-sweep wing and a predominantly titanium structure, the Boeing 2707-200 Mach 2.7 airliner was clearly the most ambitious concept. Having the reputation of the most successful developer of jetliners, Boeing was considered to be capable of solving the foreseen problems of the 2707 program and became the winner of the competition. However, after millions of dollars were spent on advanced development it was concluded that problems with empty weight, load and balance, and aero-elasticity were insurmountable.

Figure 1.3 Configuration of the Boeing 2707-300. Courtesy: Boeing.

A total design re-think in 1969 resulted in the ultimate Boeing 2707-300 design (Figure 1.3) which was based on application of a fairly highly loaded cropped delta wing in combination with a horizontal tailplane. Different from the generation of lift at low speeds with strong leading edge vortices at Concorde's highly-swept wing, Boeing preferred the 2707 wing lift to be augmented by hinged flaps at the moderate leading edge sweep. The 2707 was an extremely challenging project that never reached the prototype stage as a consequence of the US government program termination in 1971. Among the principal factors that led to this decision were concerns about the possible noise and pollution impacts of SST type aircraft:

- Many countries outlawed supersonic flight over land because of the sonic boom, which would severely restrict the projected market penetration.
- Atmospheric scientists predicted catastrophic depletion of stratospheric ozone from engine emissions, severely limiting fleet size.
- Aircraft regulators wanted the engines designed for supersonic flight to meet subsonic noise certification standards.
- Health officials were concerned about the effects of high-altitude radiation of galactic or solar origin after their observation that, at typical SST cruise altitudes between 15,000 m and 18,000 m, the radiation dose increased to double that of a subsonic jetliner cruising at 10,500 m altitude.

Others held the opinion that economic disadvantages and reordering of US national priorities were the major causes for the cancellation of the SST program. Meanwhile, a new generation of very large transonic airliners was under development in the USA and in fact many considered the Boeing 747 as a direct (in-house) competitor of the 2707.

1.3 Transonic Transport Configuration Studies

The history of near-sonic cruise airliner designs dates back to the late 1950s. One of the concepts discussed by the British STAC was the M-wing layout depicted in Figure 1.4, which was considered as an alternative to the slender wing. This novel configuration was primarily aimed at allowing cruise speeds near Mach 1.2 over land without producing a sonic boom. The M-wing incorporated highly swept thin wing segments with 45° forward sweep inboard and 45° aft sweep outboard in combination with an area-ruled fuselage[1]. Other design aspects were aimed at avoiding the poor aerodynamic efficiency and flying qualities at low speeds of a highly sweptback wing. The unusual inboard forward sweep was intended to compensate for the outboard sweep and the relatively high aspect ratio should contribute to avoiding the high vortex-induced drag of a slender wing. The STAC rejected the M-wing concept since the arguments in favor of a more ambitious Mach 2.0 cruise speed that dominated in the decision-making process. Renewed interest in the development of transonic transport began in the mid 1960s when Boeing and Lockheed generated a series of study layouts based on highly swept wings and area-ruled fuselages. These concepts complied with the principles of transonic flight successfully applied to fighters designed in the 1950s and the technology of supercritical wing sections developed at NASA-Langley. It was also realized that a transport aircraft flying at Mach 1.12 in the standard atmosphere could fly without producing a sonic boom at ground level. Since wind and non-standard temperatures change the boomless cruise speeds between Mach 1.05 and 1.25, a typical cruise speed for transonic flight is Mach 1.20. However, the irregular floor plan due to the mid-cabin body waist made it difficult to configure the cabin according to the manner that individual customers would like, and thus formed an enduring drawback of this airplane concept.

By the early 1970s it was recognized that the higher fuel prices and risk of a transonic airplane development outweighed its potential benefits, an opinion that was widely held throughout the mid-1990s. Around the year 2000 Boeing marketed a concept that was designed for extended ranges greater than 17,000 km, flying at cruise speeds of Mach 0.95 or above. It was derived from "slowing down" supersonic configurations rather than "speeding up" conventional subsonic

---

1 The transonic area rule describes how the variation of the cross-sectional area along the longitudinal axis can be manipulated to reduce the wave drag of a flight vehicle at near-sonic and supersonic flight speeds. The results of this design methodology are often manifest in highly swept wings and "coke bottle" shaped fuselages.

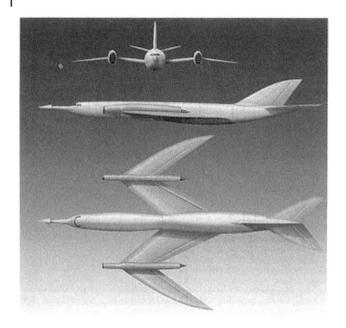

Figure 1.4 The M-wing layout for cruising at Mach 1.20, generated by the STAC in 1956.

configurations and became known as the "Sonic Cruiser". This project came to an end after the events of September 2001, when airlines that were enthusiastic about the Sonic Cruiser initially were struggling for their survival.

1.4 US High Speed Research and Development Programs

During the 1970s and 1980s several projects of the American industry were aimed at investigating applications of NASA research of advanced supersonic configurations. Study projects were part of the supersonic cruise aircraft research (SCAR) program, focusing on a second generation of supersonic airliners transporting some 300 passengers over trans-Pacific routes at speeds up to Mach 2.70. The SCAR Program was brought to an end by the marginal performance and economic potentials that appeared possible with the then available technology base. A resurgence of interest in a second-generation high-speed commercial transport (HSCT) occurred during the 1990s in Europe, the USA and Asia. Projections in 1989 for the 1995–2015 period indicated that the market in terms of passenger miles would increase by a factor of six (relative to 1971–1989) in the North-Mid Pacific and by a factor of seven in the Far East. Based on these projections, a potential market for approximately a thousand HSCT aircraft was foreseen in 1989, well over the minimum needed for a a profitable development program launch. NASA studies concluded that a supersonic transport launched in the early 21st century could be compatible with current airports, use jet fuel, and be within ten to fifteen years' technology reach.

In 1989 NASA and the US industry began investigating the potential of HSCT specifications and required technologies. The original SST of the 1960s was planned for Mach 2.70 but the required titanium structure was too heavy, and the HSCT program of Boeing and McDonnell Douglas converged on a more modest Mach 2.40, 300 seat, 9,270 km range jet A fueled aircraft as a focus for technology development. The challenges facing the HSR program were the extremely restrictive constraints placed on emissions, airfield noise, and operation costs. After approximately five years of research it was concluded that insufficient advancement in technology was available to achieve economic viability and to comply with environmental requirements. In particular an acceptable level of the sonic boom could not be achieved and the program was terminated in 1998.

1.5 European Supersonic Research Program

Similar to the US studies during the 1990s, the European industry indicated a market potential for an aircraft substantially larger and with longer longer range than the Concorde, linking the world's major cities. In 1990 the companies Aerospatiale, British Aerospace, and Deutsche Airbus launched a three-year study into the technical feasibility of a second-generation supersonic transport successor of the Concorde. In 1994 the Supersonic Research Program (ESRP) was established to undertake the research and technology development required to produce the enabling technologies for second generation supersonic commercial transport. The ESRP was supported by a common reference configuration known as the European Supersonic Commercial Transport (ESCT). Its main characteristics are compared with those of the Concorde and the Tu 144 in Table 1.1.

Also similar to the US studies during the 1990s, the European industry indicated a market potential for an aircraft substantially larger and with longer range than the Concorde, linking the world's major cities. The ESCT could be economically viable and environmentally friendly, in particular due to its capacity to carry 250 passengers over distances up to 10,000 km and its much improved take-off field performance compared to the Concorde. Figure 1.5 depicts a three view drawing of one of the designs studied in the framework of the ESCT.

Due to the widely divergent requirements at supersonic and low-subsonic flight conditions it is unavoidable that the engines for the ESCT will have a variable geometry and/or operating cycle. Figure 1.6 depicts the mid tandem fan (MTF) power plant selected for the ESCT, generated by Roll-Royce and SNECMA

Table 1.1 Characteristics of the first generation supersonic transport and the ESCT

		Concorde	Tu-144	ESCT
Maximum take-off mass	tonnes	185	200	320
Range	km	6,200	3,500	10,000
Span	m	25.6	28.8	42.0
Length	m	61.7	65.7	89.0
Passengers		90	150	250
Supersonic cruise Mach number		2.0	2.35	2.0

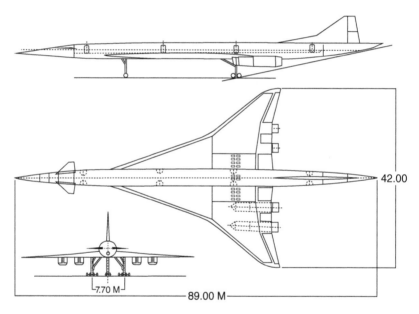

Figure 1.5 Design study of the European Supersonic Commercial Transport.

in cooperation. This engine concept is equipped with a secondary fan coupled to the secondary body. During take-off and climb the air enters the engine via auxiliary inlets. This double flow path allows very low specific fuel consumption during subsonic operation with a bypass ratio of 12 at Mach 0.8 and an exhaust velocity less than 400 m s^{-1} at the converging nozzle outlet. Auxiliary inlets are closed during the supersonic cruise at Mach 1.6 and the variable inlet mid-fan guide vanes reduce frontal airflow to the bypass duct. The bypass ratio is then 2.5 and the exhaust jet velocity 620 m s^{-1}.

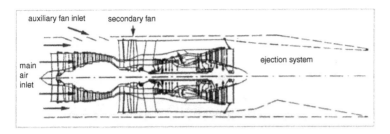

Figure 1.6 Single spool MTF in operating mode for take-off (top) and cruise (bottom).

1.6 A Market for a Supersonic Commercial Aircraft?

Ever since jet-powered airliners made their introduction into service during the 1950s, passengers on medium to long range routes have been transported at cruising speeds up to 900 km h^{-1} (Mach 0.85) in the stratosphere. Military aircraft have been able to pass the so-called "sound barrier" in routine flights since about 1960. A few exceptional types achieved continuous speeds higher than Mach 3 at altitudes above 20 km. It is therefore not surprising that after 1975 the development of a second generation supersonic airliner became a challenge to the aeronautical community. Since then, a huge amount of money has been spent on R&D programs aimed at developing advanced technology for a new generation of HSCT aircraft. *Arguably it is stunning that, despite 27 years of Concorde's satisfactory passenger service and so many technological advancements applied in all sectors of civil aviation, none of these programs have resulted in a viable development project for the near future aimed at producing an advanced supersonic commercial aircraft.*

1.6.1 Why Fly Supersonically?

Although wide-body seating during long-distance flights of a long-range subsonic airliner offers high spatial comfort, the high-priced tickets of first class and business class seating do not compensate in the form of significantly reduced boarding and traveling times. The essential economic issue is the air traveler's value of time. Some SST economic studies base the value of time on the actual earning rate for business travel and on one half the earning rate for personal travel. Concorde's concept of flying at Mach 2.0 across the Atlantic was a technical success and high-speed flying has remained attractive, especially to hasty officials.

Concordes were flagship aircraft flying at premium fares giving prestige to their passengers and operators. However, its substantial operating costs made high fares necessary: in the year 2000 the return ticket price London–New York was roughly 10,000 US dollars compared to 8,000 dollars for first class and 5,000 dollars for business class tickets of subsonic airliners. Nevertheless, Concorde's relatively high load factors and the fact that the ticket prices at the turn of the century were increasing by approximately 15% per year showed that a niche market existed for much faster passenger transport than any subsonic airliner can offer. It seems fair to assume that today a significant percentage of airline passengers is prepared to pay a premium fare, making this type of executive traveling commercially attractive to airlines. The unique achievements of the Concorde program justified sustained supersonic cruising from the technical viewpoint during its lifetime. Although technology has progressed steadily since Concorde was conceived, it was decades ahead of its time and nowadays we cannot do significantly better. Nevertheless, new technical innovations and organizational approaches will be mandatory to develop and operate a second generation SCT in the economic and regulatory environment of the 21st century.

Having surveyed the abundance of research achievements and project proposals generated during the half century after Concorde's first flight, one could anticipate that significantly improved concepts have become available in most aeronautical disciplines and production capabilities that could lead to a realistic program for development, production, and operation of an environmentally acceptable and economically viable second generation supersonic airliner. A crucial condition for such a program is that a new HSCT will be developed and produced by a consortium of R&D institutes and companies in America, Europe, and East-Asia. Since all engineers involved in the first generation supersonic airliners are no longer available to apply their knowledge to such a development, considerable effort will be required to bring together and educate sufficiently experienced staff. The availability of relevant progress reports of previous projects will be indispensable to make such an international project team manageable and effective.

1.6.2 Requirements and Operations

Arguments in favor of developing and producing a modernized version of the Concorde would not immediately get acclaim from airlines. In the present commercial aviation market its 110 passenger cabin would be too small, its transatlantic design range too short, its fuel economy too low, and its engines too noisy when taking off. Although Concorde's technical complexity made it a very costly aircraft to purchase, its high operating costs were associated primarily with its poor fuel efficiency, high maintenance, and upgrading costs.

A new high-speed transport aircraft would fly over the Atlantic, the Pacific, and uninhabited areas, covering about 80% of the most attractive long-range routes where supersonic flight is legally permitted. The size of the market, estimated as being between 500 and 1,000 aircraft, suggests that there will only be room for a single development program and only international cooperation would make such a program feasible. Enabling a potential trip time reduction of 50% or more when compared to current subsonic flights, supersonic air travel is the one technology that offers a large step forward in functional capability and a large increase in service. This increased productivity potential could result in SCT that is economically viable and environmentally acceptable and thereby could capture a significant portion of the long-range travel market.

Since an SCT will have to comply with the same international regulations as the contemporary subsonic fleets, take-off performance and engine design must be improved considerably relative to Concorde's capabilities. Cruise speed is a major factor affecting the operating costs and it is the primary performance characteristic that has to be considered in drawing up the top level specifications, and its choice has far-reaching consequences for the design and development as well as the operation of the aircraft.

- The Boeing 2707-200 was designed to achieve a range of 6,600 km, similar to the trans-Atlantic routes served by Concorde. Such a maximum range would be of limited interest for the market of a future SCT since the most important part of its market will be the long distances over water, in particular the trans-Pacific routes with ranges of more than 10,000 km.
- The SCT must be able to take-off from and land on existing airfields and comply with the associated noise criteria applicable to present-day jetliners and the plane's dimensions must be compatible with the existing infrastructure of the relevant airports. Accordingly, the accessibility to the aircraft must allow for parallel embark and disembark, service, and fueling in order to enable rapid turn-around.
- In order to serve the many routes that have overland legs, subsonic/transonic flight performance must be at least as good as supersonic cruising and the plane should be able to cruise at speeds up to Mach 1.2 without producing an offensive sonic boom, thereby enabling increasing the cruise speed over land by 50% relative to present-day jetliners.

1.6.3 Block Speed, Productivity, and Complexity

- The block time for intercontinental supersonic flight rapidly improves through the low Mach number region; it levels out at speeds above Mach 3.0. Greater speeds will not be paid off with appreciable time saving to the passenger as well as increased productivity to the airliner, and the cost of cruising faster than Mach 2.0 can be large since it complicates the airframe and systems development effort. In particular, the structure of a high Mach number aircraft is subject to kinetic heating of the airframe skin. This requires a complicated air conditioning system and the usage of expensive heat-resisting structural materials, whereas the combination of materials having different coefficients of expansion may increase structural stresses.

- Complicated variable-geometry engines are required when flying at high Mach numbers and, since the best cruise altitude increases as well, the installed power plant becomes heavier and more costly. Moreover, a heavier fuselage structure is required to cope with the higher cabin pressure differential and increased fuel tank pressurization to prevent fuel boil-off.
- A cruise speed lower than Mach 2.0 leads to less wing sweep than Concorde's 60° leading edge sweep, which is better suited to low speed operation, higher bypass ratio engines that reduce take-off noise, and cruise altitudes that reduce global impact of emissions. A cruise speed of Mach 1.6 to Mach 1.8 offers a practical possibility for increasing the block speed to about twice that of present-day jetliners.

These considerations demonstrate that a considerable development effort is required to combine the need for high fuel efficiency in supersonic cruising flight with acceptable development costs and friendliness to the airfield environment during take-off, climb-away, approach, and landing. This means a major dilemma for the design team of any SST: there is a fundamental discrepancy between design characteristics acting in favor of efficient high-speed cruising and acceptable flight characteristics at subsonic speeds, in particular take-off and landing. A solution may be immanent in a market analysis indicating the effect of increasing the block speed on the aircraft's productivity and economy on a particular route network.

The industrial activities aimed at development of new SCT applications were concentrated in the time frame 1960–1990 but, in spite of the long history of technological research and development on civil supersonic aircraft, little systematic information required to initiate a realistic conceptual design of a supersonic transport or executive jet has been published. Remarkable exceptions are Corning's textbook [2] appearing first in 1960 with later versions up to 1976, and [3] published in 1978. Küchemann's authoritative book is dedicated to the aerodynamic design of transport aircraft in general and Concorde's aerodynamic development in particular.

The Concorde would not be able to successfully comply with the requirements of commercial air transport in the 21st century but with the present-day technologies a much more efficient supersonic transport than the Concorde could be built. Many projects have been started to investigate the viability of a second generation SCT, resulting in a wealth of articles written by investigators from all continents, together forming a deluge for (teams of) engineers who are supposed to create a design concept based on a realistic set of top level requirements. It is the intention of the present author to present a synthesis of classical analysis models as well as methodologies generated by recent technological research and project studies that can be considered as an essential guidance to conceive an initial configuration design.

Bibliography

1 Blackall, T.E. *Concorde, the Story, the Facts, and the Figures* Foulis & Co., Ltd; 1969.

2 Corning, G. *Supersonic and Subsonic, CTOL and VTOL, Airplane Design.* 4th ed. College Park, MD: University of Maryland; 1976.

3 Küchemann, D. *The Aerodynamic Design of Aircraft*, 1st ed. Oxford: Pergamon Press; 1978.

4 Torenbeek, E., and Wittenberg H. *Flight Physics – Essentials of Aeronautical Disciplines and Technology, with Historical Notes.* Springer; 2009.

5 Brandt, S.A., Stiles R.J., Bertin J.J., and Whitford R. *Introduction to Aeronautics: A Design Perspective*, AIAA Education Series. Washington, DC: AIAA Inc.; 1997.

6 Anderson Jr, J.D. *The Airplane; A History of Its Technology.* Reston, VA: American Institute of Aeronautics and Astronautics; 2002.

7 Raymer, D.P. *Aircraft Design: A Conceptual Approach/.* 4th ed. AIAA Education Series. Reston, VA: AIAA Inc.; 2006.

8 Morgan, M.B. Supersonic Aircraft – Promise and Problems. *J. R. Aeronautical Soc.*, June 1960, 64(594):315–334.

9 Küchemann, D. *Aircraft Shapes and Their Aerodynamics for Flight at Supersonic Speeds.* Pergamon Press; 1962.

10 Maurin E., Vallat P., Harpur N.F. Struktureller Aufbau des Überschallverkehrsflugzeuges "Concorde". Luftfahrttechnik und Raumfahrttechnik. 1966, January, 12.

11 Swan, W.C. A Review of the Configuration Development of the US Supersonic Transport, Paper 17. 11th Anglo-American Aeronautical Conference, London, UK, September 8–12; 1969.

12 Swihart, J.M. The Promise of the Supersonics. AIAA Paper No. 70-1217. 6th Propulsion Joint Specialist Conference, June 15–19, 1970, San Diego, CA, USA; 1970.

13 Morien, Sir Morgan. A New Shape in the Sky. Aeronautical J., January, 1972.

14 Swan, W.C. Design Evolution of the Boeing 2707-300 Supersonic Transport. Part I, Configuration Development, Aerodynamics, and Structures AGARD CP 147, October, 1973.

15 Poisson-Quinton, Ph. First Generation Supersonic Transports. ONERA TP 1976-113, 1976.

16 Shevell, R.S. The Technical Development of Transport Aircraft – Past and Future. AIAA Paper No. 78-1530, August 1978. https://doi.org/10.2514/3.57876

17 Forestier, J., Lecomte P., and Poisson-Quinton Ph. Les Programmes de Transport Supersonique dans les Années Soixante'. Proceedings of the European Symposium on Future Supersonic/Hypersonic Transportation Systems, Strasbourg, November, 1989.

18 Reimers, H.D. Das Überschallverkehsflugzeug der Zweite Generation – Eine Zweite Chance?! DGLR Jahrbuch, 93-03-029:1239–1250; 1993.

19 Seebass, R., and Woodhull J.R. History and Economics of, and Prospects for, Commercial Supersonic Transport. RTO AVT Course on Fluid Dynamic Research on Supersonic Aircraft, Rhode-Saint-Genèse, Belgium, Published in RTO EN-4, 25–29 May, 1998.

20 Collard, D. Concorde Airframe Design and Development. SAE Trans. 100:2620–2641; 1991. www.jstor.org/stable/44548119.

21 Mercure, R.A. NASA's Supersonic Commercial Aircraft Technology Development – Background and Current Status. ICAS Congress Presentation, September 2002.

22 Torenbeek, E., Jesse E., and Laban M. Conceptual Design and Analysis of a Mach 1.6 Airliner. AIAA Paper No. 2004-4541, September 2004. https://doi.org/10.2514/6.2004-4541

23 Mathieu, M.S., et al. Preliminary Design of a N+1 Overwater Supersonic Commercial Transport Aircraft, AIAA Paper 2017-1387. https://doi.org/10.2514/6.2017-1387

2

The Challenges of High-speed Flight

If the history of flight has shown us anything, it has shown us that aeronautics has always been paced by the concept of faster and higher. Although this has to be mitigated today by economically viable and environmentally safe airplanes, the overall march of progress in aeronautics will continue to be faster and higher.

John D. Anderson Jr. (2002)

The following global overview will describe several opportunities for starting a revival of the challenging branch of aeronautics that came to an unfortunate end with the cancellation of Concorde's operations in 2003. Using information from Concorde's development and recent design projects, a set of initial requirements for a next generation SCT will be proposed. In particular, it will be a crucial factor for the feasibility of any SCT development program that it is conceived and produced by a consortium of industrial participants and governmental institutes in the United States, Europe, and Asia.

Although the unique achievements of the Concorde program justified sustained supersonic cruising from the technical and operational viewpoints during its lifetime (1976–2003), new organizational approaches will be mandatory to develop, produce, and operate second generation supersonic commercial transport (SCT) in the economic and regulatory environment of the twenty-first century. Different from the early days in which the Concorde and the American SST projects were developed, the design team will have to justify the reasons why a second SCT generation is worthy of consideration. The new SCT will have to comply with all contemporary certification rules regarding safety, noise, and pollution, and have the same level of reliability and operational costs. And it must be developed and produced economically, without undue technical risks and in sufficient numbers to become profitable. Having surveyed an abundance of research achievements and project studies generated after the introduction of the Concorde, the present

Essentials of Supersonic Commercial Aircraft Conceptual Design, First Edition. Egbert Torenbeek.
© 2020 Egbert Torenbeek. Published 2020 by John Wiley & Sons Ltd.

author has embraced the scenario that the present technological state-of-the-art is sufficiently mature, and economical conditions are favorable, to develop and produce a safe and economically viable second generation supersonic airliner.

2.1 Top Level Requirements (TLR)

In order to avoid a waste of time and money spent in an unfeasible undertaking, effort must be made to gain an understanding of the operational characteristics to be incorporated in the vehicle as well as the dominant needs and conditions having a major effect on the problems that have to be tackled by the SCT advanced design team. An essential activity to start an advanced design project should be drawing on the top level requirements, which should not be based on approaches inspired by unrealistic expectations based on out-of-the-box brainstorming. In particular, the most daunting challenges facing the American technology exploration programs abandoned in 1971 and 1999 were the aggressive Mach 2.7 cruise speed of the SST project of the 1960s and those of the HSCT program of the 1990s, with very restrictive constraints on emissions, noise, and operating costs.

Arguments in favor of developing and producing a modernized version of the Concorde would not immediately get acclaim from airlines. In the present commercial aviation market its 100 passenger cabin would be too small, its transatlantic range too short, and its engines too noisy during and immediately after the take-off. Concorde was a very costly aircraft: nine were sold in the 1970s to British Airways and Air France for a price of 80 million US dollars each. Concorde's high operating costs were associated primarily with the aircraft's low fuel efficiency and the high maintenance and upgrading costs associated with its technical complexity. The small size and high costs of the vehicle confined it to a very small ultra-premium market.

The Asia/Pacific rim market is considered to be a most important element in the planning for the supersonic market. Since most experts do not believe that it will be possible to reduce the sonic boom over land to acceptable levels, the HSCT must survive with supersonic flying over water. Compared to Concorde's performance it must have much better flight efficiency in off-design conditions such as high-subsonic flight, whereas its take-off and landing performance must comply with the same environmental requirements that apply to subsonic jet airliners.

Summarizing, it is likely that the second HSCT generation will be characterized by the following specifications:

- The aircraft shall have the capacity to carry 250 mixed class passengers over a distance of 10,000 km, comprising a maximum of one supersonic cruising leg and two transonic (Mach 0.95) legs. Cruising at an airspeed of at least

1,700 km h^{-1} (Mach 1.60) will save up to five hours flying time compared with present-day subsonic airliners, making it especially attractive on trans-Pacific flights.

- It is not unlikely that in the future speeds up to Mach 1.20 will be accepted over land if the plane produces little or no susceptible sonic boom. This will enable it to increase the cruise speed by 40% relative to present-day jetliners and cruise a first leg with no performance penalty until water is reached and acceleration to cruise speed becomes possible.

- Fuel consumption in supersonic cruising flight will have to improve on Concorde's payload-fuel efficiency by a factor two at least, and in high-subsonic flight it shall be at least as good as in supersonic cruise.

These considerations indicate that a considerable development effort is required to combine the need for high fuel efficiency in supersonic cruising flight with acceptable development costs and friendliness to the airfield environment during take-off, climb-away, approach, and landing. This means a major dilemma for the design team of any SST: there is a fundamental discrepancy between design characteristics acting in favor of efficient high-speed cruising and acceptable flight characteristics at subsonic speeds, in particular during take-off and landing. A solution may be immanent in a market analysis indicating the effect of increasing the block speed on the productivity and economy on a particular route network.

2.2 The Need for Speed

Since cruise speed is an essential factor affecting the commercial productivity as well as operating costs, the efficiency of air travel is (and has always been) closely related to speed. The cruise Mach number is therefore the primary performance characteristic to be considered in the design process of drawing up the TLRs. Experience with the Concorde has confirmed that passenger comfort is enhanced considerably by the reduction in flying time compared to its subsonic counterparts. Passengers tend to become tired after approximately four hours of flying and cruising at supersonic speed makes a transatlantic flight much more convenient compared to the six to seven hours flight of a high-subsonic jetliner.

The block time of a high-speed flight is not inversely proportional to the cruise Mach number since supersonic transport needs at least one hour extra time for take-off, climb, transonic acceleration, subsonic descent, and landing (Figure 2.1), whereas a subsonic airliner needs roughly one half hour. For example, a 50% increase in cruise Mach number from Mach 2 to Mach 3 on a 6,500 km flight from London to New York effectively saves only about thirty minutes, and the saving in traveling time is reduced even more when long subsonic cruise legs have to be flown overland due to sonic boom limitations. High supersonic cruise

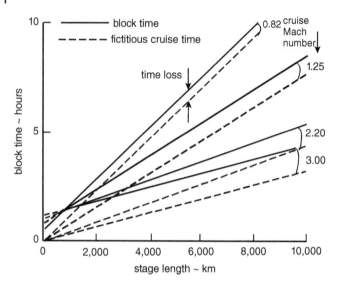

Figure 2.1 Reduction in traveling time with increasing cruise Mach number.

Mach numbers will therefore pay off only on the very long distances typical of trans-Pacific routes. For instance, at Mach 2.0 cruising speed the flight time between London and Tokyo could be reduced from thirteen to seven hours.

For airline operators the average number of flights that can be flown on one day is crucial to achieve a high utilization. Two flights can be made on typical trans-atlantic routes cruising at Mach 0.8. This increases to three flights for $M > 1.2$ and to four flights at $M > 2.0$. The productivity of an SCT is therefore high compared to subsonic transports on the provision that the time losses on the ground due to unreliability do not increase. Since this principle holds as well for airlines with a mixed route structure, the design speed and size of the new SCT must rely on realistic traffic predictions and in-depth studies of route structures in which the cruise Mach number is an essential parameter.

2.3 Cruise Speed Selection

Selection of the cruise Mach number of a supersonic transport is widely considered as the most controversial design parameter. The history of design projects aimed at the development of SCT cruising at speeds significantly higher than Mach 2.0 has demonstrated that this step will be at least one too far. In fact, the prospects of successfully developing and operating second generation SCT appear to be significantly brighter when it is designed for a lower cruise Mach number than Concorde's Mach 2.0. Many complications associated with developing very high-speed commercial transport are related to the following topics.

- The cost of cruising faster than Mach 2.0 can be large since the airframe and the passenger cabin are subject to kinetic heating, which complicates the airframe and the air conditioning system development effort considerably. In particular, the structure of an aircraft flying at speeds in excess of Mach 2.0 is subject to serious kinetic heating of the skin, which requires the usage of expensive structural materials such as heat-resistant composites, titanium, and even ceramics.
- Although speeds higher than Mach 2 are technically realizable, one doesn't achieve as much as one would like because it takes longer to climb to the supersonic cruise condition and trade-offs have to be evaluated. For instance, a basic consideration could be that the project must survive flying supersonically over water. Since routes such as Europe–Japan have to be operated mostly over water, enormous time savings can be made compared with a large circular route flown partly over land at subsonic speed.
- For airline operators the average number of flights that can be flown on one day is crucial to achieve high utilization. Two flights can be made on transatlantic routes cruising at Mach 0.8. This increases to three flights for $M > 1.2$ and to four flights at $M > 1.8$. Compared to subsonic airliners, the productivity of SCT is therefore high on the provision that the time losses on the ground due to unreliability do not increase. Since this principle also holds for airlines with a mixed route structure, the design speed and size of the new SCT must rely on realistic traffic predictions and in-depth studies of route structures in which the cruise Mach number is treated as an important parameter.

Arguably, Mach 1.6 SCT has 15% lower productivity than a Mach 2 vehicle with the same payload and it is a challenge to demonstrate that this disadvantage can be more than compensated by a set of factors making its reduced cruise speed economically and operationally attractive. Since there exists no operational first generation SCT, second generation SCT will have to compete with long-range subsonic airliners and compared with subsonic transport the cruise speed of a Mach 1.6 airliner is twice as high. This causes the block time of a 10,000 km flight at Mach 1.6 to decrease from about twelve hours to seven hours, and the productivity to increase by 70% for the same yearly flight hours. However, the design problems of supersonic transport aircraft increase considerably with the speed regime aimed at, and the following aspects may be crucial for the feasibility of a modest cruise Mach number design.

- A high aerodynamic efficiency L/D in cruising flight is obtained only if the wing's leading edge is predominantly subsonic. For Concorde's Mach 2 cruise speed this resulted in an aspect ratio 1.6 delta wing with an average leading-edge sweep angle of 60°, resulting in a poor L/D in the take-off configuration. This necessitated four reheated Olympus turbojets, together producing a take-off thrust/weight ratio of 0.37. However, accepting a cruise speed of Mach 1.6 allows less leading-edge wing sweep than Concorde's 60°, enabling a delta or arrow wing with a typical aspect ratio up to three. This can be achieved with a relatively thick inboard wing with blunt subsonic leading edges combined with a relatively thin outboard wing with sharp supersonic leading edges. Such a wing shape yields a significantly improved aerodynamic efficiency in all flight regimes.

- Kinetic heating of the airframe skin is far less severe at Mach 1.60 than at Mach 2.0. This relaxes the problem of designing for thermal stresses in the expanding structure and reduces the environmental control system capacity, resulting in reduced fuel consumption, structural weight, and costs.
- The overall efficiency of a turbojet engine such as Concorde's Olympus improves between Mach 1.60 and Mach 2.0. However, a low bypass ratio turbofan has similar efficiency at Mach 1.60 and much better efficiency at subsonic speed than the Olympus at Mach 2.0. Moreover, a lower cruise altitude can be selected for a Mach 1.60 transport, resulting in a lower installed engine thrust, weight, and cost. These conditions reduce community noise and global impact of emissions and are better suited to improve flight efficiency ML/D during supersonic flight as well as subsonic operation over land. Moreover, the operating conditions of the engine inlet system in subsonic and supersonic cruising flight regimes are closer together, which requires less (inlet) geometry variation, reduced cost, and improved inlet efficiency.
- Complicated variable-geometry engines are required when flying at high Mach numbers above 1.5 and consequently the installed thrust increases and the power plant becomes heavier and more costly. Since the optimum cruise altitude increases as well, a heavier fuselage structure to cope with the higher cabin pressure and increased fuel tank pressurization to prevent fuel boil-off are required.

The challenges of designing supersonic SCT increase progressively with the cruise Mach number and the aerodynamic mismatch between supersonic cruising flight and off-design conditions increases. This mismatch leads to more design complications when the cruise Mach number and the sweep angle of the wing increase. Similar to high-lift devices applied on most wings of several military aircraft such as the F 111, a solution to cope with this problem is to incorporate a wing with variable leading-edge sweep angle, which is known as a swing wing, has been applied to supersonic cruise vehicles (SCVs). However, since this application of variable sweep leads to considerable complications and weight increase of the wing structure, it is worth exploring the feasibility of applying the oblique wing. This exotic configuration has been studied in several advanced design projects since the 1970s and since the results of these investigation are very promising, oblique wing aircraft configurations will be the subject of Chapter 10.

2.4 Aerodynamic Design Considerations

Acceptable supersonic aircraft designs must comply with stringent aerodynamic design criteria. In particular, the flow over the aircraft must be stable and controllable in all flight conditions and the vehicle must be stable and controllable in all flight phases, including off-design conditions associated with maneuvering and stalling. A crucial problem for the SCT design team is to generate an external shape producing a steady and controllable flow in normal flight conditions as well as off-design conditions such as take-off, transonic flight, and landing. In contrast to flight at subsonic speed, most of the wing lift of a supersonic plane is obtained from compression forces acting on the lower wing surface instead of low pressures acting on the upper surface. In this respect, the accumulated experience of SCV designers has indicated that it is a sensible global aim to keep (most of) the wing's leading edge behind the Mach waves and to keep the whole airplane well within the Mach cone from the (fuselage) nose.

When the vehicle penetrates and exceeds the transonic flow regime its aerodynamic efficiency degrades significantly and a primary objective of the aerodynamic designer is to avoid strong shock waves by applying suitable geometric principles. Typical measures are: selecting an appropriate general arrangement of the overall configuration, using smooth fuselage–wing combinations by means of area ruling, and smart integration of the power plant into the airframe.

2.4.1 Fuel and Flight Efficiency

The fuel efficiency of a passenger transport is a basic parameter for defining its flight economy. It is the product of the number of cabin seats and the distance that can be traveled by burning a specified amount of fuel, briefly expressed in terms of seat-kilometers per liter. The fuel efficiency of a high-subsonic airliner in cruising flight does not differ significantly from low speed conditions. However, the Concorde consumed almost three times the fuel required for a subsonic airliner transporting the same payload over the same distance and the fuel efficiency of an SCV in supersonic cruising flight is considerably lower than during the subsonic flight segments[1].

---

1 Some experts consider efficient supersonic flight as an oxymoron.

The (momentary) fuel efficiency in level flight is proportional to the flight efficiency $\mathcal{P} \overset{\text{def}}{=} \eta_0 L/D$, with η_0 denoting the overall power plant efficiency. The notion of flight efficiency, also known as the range parameter, is not standardized and sometimes replaced by the product ML/D. The use of ML/D instead of $\mathcal{P}$ is based on the observation that in subsonic flight the thrust-specific fuel consumption (TSFC) does not vary greatly with speed. Hence, η_0 is roughly proportional to the flight Mach number.

During the development of the Concorde, devoted proponents of supersonic transportation suggested that the fuel efficiency at supersonic speed is not very different from the fuel efficiency at subsonic speed, arguing that the deterioration of L/D caused by supersonic wave drag is compensated by the high flight Mach number of SCT. This argument was inspired by the observation that Concorde's expected cruise performance $ML/D \approx 15$ at Mach 2 will be not much different from the same figure for (contemporary) subsonic airliners [2]. The shortcoming of this reasoning is that it ignores the significantly increasing thrust specific fuel consumption (TSFC) with Mach number and hence the variation of ML/D differs considerably from the variation of $\eta_0 L/D$ with Mach number.

2.4.2 Aerodynamic Efficiency

The aerodynamic efficiency L/D in the cruise configuration of a flight vehicle of specified geometry is basically affected by the Mach number and the lift coefficient. Theoretically, there exists a unique combination of these variables resulting in maximum fuel efficiency, which we refer to as the global optimum condition for altitude and speed. In reality the flight conditions are mostly constrained by operational limits such as engine ratings, Mach number, and altitude constraints [4]. Although advanced designers seem to have a lot of possibilities to improve the maximum aerodynamic efficiency of their design relative to previous aircraft, their freedom is constrained by available design technology and economic considerations. An educated guess of what is achievable must be therefore be made in the conceptual design stage of an SCV by collecting literature data on maximum L/D figures quoted in the literature for existing transport aircraft or validated project information.

As an example, Figure 2.2 illustrates that subsonic high-capacity jet transport in cruising flight designed around the year 2000 achieves a maximum L/D between approximately 18 and 20, whereas the next generation of long-range subsonic airliners is predicted to achieve an aerodynamic efficiency up to 20. At transonic Mach numbers the aerodynamic drag exhibits a progressive drag rise constraining the speed of jetliners to less than Mach 0.95, typically. A supersonic airliner in subsonic flight exhibits an aerodynamic efficiency at least 20% less than contemporary subsonic airliners, unless it incorporates variable wing geometry. The

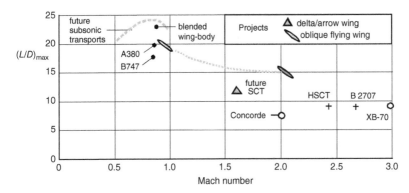

Figure 2.2 Aerodynamic efficiency versus Mach number of subsonic and supersonic commercial transport aircraft (projects).

transonic drag rise is, however, not very large; the achievable L/D decays gradually at speeds above Mach 1.2, and aerodynamic efficiencies between nine and fifteen are predicted for a next generation HSCT in supersonic cruising flight.

2.4.3 Power Plant Efficiency

The overall efficiency of gas turbine engines designed for civil aircraft propulsion is the product of the basic system process efficiencies: inlet efficiency, thermal efficiency of the core engine, combustion efficiency, efficiency of the power transfer to the fan, and the propulsive efficiency.

- The attainable thermal efficiency of a modern turbofan at subsonic speeds is approximately 45%, with a target of 50% in the year 2020. The product of the other efficiencies may increase at high-subsonic speeds to 85% for high bypass ratios. Figure 2.3 illustrates that the overall power plant efficiency at low subsonic speeds increases more or less proportionally to the Mach number, indicating that the TSFC is nearly constant. This trend levels off at approximately Mach 0.5 and the overall efficiency no longer increases at transonic speeds. The presently achievable overall efficiency in high-subsonic flight is close to 45%.
- The overall efficiency of turbojets and turbofans designed for HSCT application increase due to the progressively increasing total inlet pressure, an effect that levels out between Mach 2 and Mach 3. The points in Figure 2.3 indicate that Concorde's Olympus engine was unique with its installed overall efficiency $\eta_0 \approx 0.40$. However, its overall efficiency at subsonic speeds was very poor, resulting in a large fuel consumption during subsonic cruising and diversion.

Next generation HSCT will probably need low bypass ratio turbofans with overall efficiencies of at least 35% and 42% at subsonic and supersonic Mach numbers,

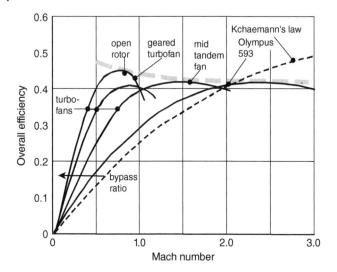

Figure 2.3 Overall power-plant efficiency of commercial transport.

respectively. Consequently, the propulsive efficiency of a high-speed vehicle during supersonic flight can be superior to that of a subsonic airliner. Engines designed for efficient propulsion at both speed regimes will, however, be very complex since they feature a variable flow cycle as well as variable-geometry inlet and exhaust systems.

2.4.4 Flight Efficiency

Design studies of projected SCTs quote maximum L/D ratios 40% higher than Concorde's aerodynamic efficiency in cruising flight. However, achieving a higher value should be the prime objective of any future SCT design effort. For example, it is noticeable that viscous drag in supersonic flight is of similar magnitude as wave drag and induced drag. In spite of this, many publications treating supersonic flow analysis and design optimization ignore the possibility of reducing viscous drag altogether – a simplification that is not justified in realistic SCT design synthesis.

On the basis of Figure 2.4 it is concluded that Concorde's flight efficiency in cruising flight approached $P = 3$, whereas in subsonic flight it did not exceed 2. In contrast, the maximum flight efficiency in cruising flight for present-day jetliners varies between 6 and 8. Consequently, future second-generation HSCT should

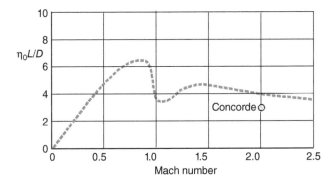

Figure 2.4 Potentially achievable flight efficiency of supersonic transport.

attain an average flight efficiency in cruising flight between at least 4 and 5, with no serious deterioration at high-subsonic speeds.

2.4.5 Cruise Altitude

The proper (initial) choice of a cruise altitude should be based on an optimization procedure with a solution depending on the design stage – a quasi-analytical approach to this design problem is proposed in Chapter 8. An optimum condition for the aircraft's wing loading and the initial cruise altitude can be derived by assuming that the engines are sized to balance the drag in cruising flight and then deriving the condition for minimum take-off gross weight. Even for a fully developed advanced design of SCT the computation of its best cruising altitude can be a complicated process.

In brief, the optimum cruise altitude for an aircraft designed to fly at a specified Mach number is below the altitude where maximum L/D occurs. Although the fuel consumption increases below the altitude for minimum drag, the required engine size and its installation weight decrease. The altitude where these effects compensate each other can be considered as the optimum cruise altitude, resulting in minimum take-off weight, whereas it appears that the best cruise altitude of a supersonic transport increases when its design Mach number increases. For instance, the fuel-economic altitude typically increases from 14,000 m for Mach 1.6 to 17,000 m for Mach 2, whereas the best cruise altitude of a Mach 3 aircraft is more than 20,000 m above SL. Obviously, the consequences for the power plant and its installation are daunting and the same applies to the pressure structure and the environmental control system.

Bibliography

1 Corning, G. *Supersonic and Subsonic, CTOL and VTOL, Airplane Design.* 4th. ed. College Park, MD: University of Maryland; 1976.

2 Küchemann, D. *The Aerodynamic Design of Aircraft*, 1st ed. Oxford: Pergamon Press; 1978.

3 Sobieczky, H. (ed). *New Design Concepts for High Speed Air Transport*, Springer-Verlag Wien GmbH; 1997.

4 Young, T., *Performance of the Jet Airliner.* Wiley and Sons; 2017.

5 Nicholson, L.F. Supersonic Transports – Some Considerations of Non-Cruising Problems, ICAS Paper No. 51, August 1962.

6 Wilde, M.G. Supersonic Transport Aircraft. *J. R. Aeronautical Soc.* 65 (602):75–0110; 1961.

7 Küchemann, D., and Weber J. An Analysis of Some Performance Aspects of Various Types of Aircraft Designed to Fly over Different Ranges at Different Speeds. *Progr. Aeronautical Sci.* 9:324–456; 1968.

8 Swan, W.C. A Review of the Configuration Development 0f the US Supersonic Transport", 11th. Anglo-American Conference, September 8–12, 1969.

9 Morgan M. A New Shape in the Sky. *Aeronautical J.* January:1–18; 1972.

10 Kulfan, R.M. *High Speed Civil Transport Opportunities, Challenges and Technology Needs.* Trans. *Aeronautical Astronautical Soc.* Republic of China. 25(1):1–20; 1993.

11 Mertens, J. Son of Concorde, a Technology Challenge. Ch. 3 of [3].

12 Kroo, I. Unconventional Configurations for Efficient Supersonic Flight. von Kármán Institute for Fluid Dynamics, Lecture Series 2005-06, June 6–10, 2005.

3

Weight Prediction, Optimization, and Energy Efficiency

Over the many decades of commercial airplane development designers have focused on minimizing the maximum take-off weight. More recently, the progressively increasing fuel prices and environmental constraints have shifted the emphasis to reducing fuel burn-off. In view of the large fuel weight fraction of a high-speed airliner, reducing the mission fuel has to get even more emphasis than for a subsonic airliner. However, weight engineering is a highly iterative process, and an accurate weight breakdown is not available until the conceptual design phase is completed. Reference is made to [4] for a discussion on the various types of weight prediction methods and an explanation of a typical methodology that can be used for early prediction of the major weight components. The following text is primarily intended to demonstrate the sensitivity of the airplane's all-up weight (AUW) to the flight efficiency, thereby emphasizing the importance of the lowest possible drag and high propulsive efficiency in cruising flight.

3.1 The Unity Equation

The separation of basic functions for classical airplane configurations allows us to decompose the AUW of the aircraft when taking off for flying the design range with the design payload, using the following symbols:

W_{to} – maximum take-off weight (MTOW)
W_{oe} – operating empty weight (OEW)
W_{pay} – design payload (DPL)
W_{fuel} – block fuel weight (TFW).

The following expression, stating that the sum of all basic weight fractions contributing to the MTOW equals one, is known as the "unity equation":

$$\frac{W_{oe}}{W_{to}} + \frac{W_{pay}}{W_{to}} + \frac{W_{fuel}}{W_{to}} = 1. \tag{3.1}$$

Essentials of Supersonic Commercial Aircraft Conceptual Design, First Edition. Egbert Torenbeek.
© 2020 Egbert Torenbeek. Published 2020 by John Wiley & Sons Ltd.

Depending on the availability of design information, the unity equation can be manipulated in various ways by further breaking down the three basic weight components. Using Equation (3.1) offers the advantage that the sensitivity of the weight distribution to the most influential design characteristics is obtainable without resorting to cumbersome iterative design cycles.

The present approach was originally developed for subsonic jetliners but the basic method is applicable to supersonic aircraft as well, on the provision that various weight terms have been calibrated with data of existing or projected aircraft in the same category of supersonic designs. The prediction has class I accuracy and can be used for the early design cycle of supersonic transport as well as a supersonic executive aircraft. For example, application to the Concorde weight distribution was successful mainly due to the availability of data on the aerodynamic and the overall propulsive efficiencies in cruising flight. In order to make a credible assessment of leading technological parameters, these efficiencies can be inserted as independent design variables.

3.2 Early Weight Prediction

The MTOW of a commercial transport is typically determined by the design mission range R_d that can be traveled with the design payload W_{pay}. The mission range is usually determined by the number of passenger seats to be installed in a cabin in mixed class seat arrangement. Alternatively, the MTOW may be derived from the distance to be flown with the payload in high-density seating, the harmonic range R_h. Although flying the design mission range requires significantly more fuel than flying the harmonic range, the airplane takes off with the same MTOW for both missions. This process may require two design cycles, resulting in different weight breakdowns.

3.2.1 Empty Weight

The empty weight fraction of subsonic airliners varies roughly between 0.65 for short-range aircraft and 0.45 for long-range aircraft. An empty weight prediction based on statistics can be made as soon as the fuel weight fraction is known [4]. However, early empty weight estimations of an SCT generation may suffer from inaccuracies of at least 10%, resulting in considerable problems during the more advanced stages of the development. An extreme example of weight growth during the design development phase is the Concorde, whose MTOW increased from about 60,000 kg in the early design stage to 185,000 kg for the final production version. Remarkably, Concorde's empty weight fraction turned out to be only 0.44, a statistically normal percentage regarding its large total fuel weight of approximately 50% of the MTOW.

The decomposition of the OEW is based on the assumption that the DPL, the MTOW, and the installed take-off thrust T_{to} are the principal components affecting the empty weight as follows:

$$W_{oe} = C_{sys}(C_{pay}W_{pay} + C_{af}W_{to}) + C_{pp}T_{to} + C_{sys} + W_{fix}, \qquad (3.2)$$

including the following three components:

- The first bracketed term, denoted as the body group, represents the fuselage weight, which is determined primarily by the fuselage dimensions and the maximum number of seats in the passenger cabin. Other components belonging to this category are the weight of cabin furnishings and equipment, and operator items. Since the vertical tail size is determined primarily by the fuselage geometry and the cabin configuration, its structure weight is also classified as a body group weight component[1]. The factor C_{pay} for subsonic single-deck airliner fuselages depends primarily on the number of seats abreast in the main cross-section. Typical (approximate) values for narrow- and wide-body subsonic transport are $C_{pay} = 1.25$ and $C_{pay} = 1.50$, respectively.
- The second term – referred to as the airframe weight – summarizes components that are sized mainly to the MTOW defining the critical loading condition, W_{af}. The wing structure and the landing gear typically belong to this category since their weights are considered functionally and statistically proportional to the MTOW. The wing structure weight fraction of subsonic airliners varies typically between 0.09 and 0.13, and if a horizontal tail or a fore-plane is present its weight can be assumed to be proportional to the wing structure weight. Concorde's wing structure weight fraction of not more than 0.074 can be ascribed to the low aspect ratio, high inertia relief due to fuel, and the absence of complex high-lift devices. Similar to subsonic transports, Concorde's twin-leg main landing gear has a weight fraction of 0.039. However, due to the plane's high thrust/weight ratio, its power plant weight fraction of 0.13 is much higher than the typical 0.08 for subsonic jetliners.
  Statistics indicate that for present-day subsonic jetliners C_{af} is between 0.20 and 0.22. Taking into account that Concorde was designed during the 1960s, its airframe weight fraction $C_{af} \approx 0.24$ is considered to be fairly low.
- The third term represents the installed power plant weight, which amounts to typically 30% of the total installed thrust; hence $C_{pp} \approx 0.125$.

---

1 The Concorde's (large) aluminum fin weight amounts to 28% of the fuselage structure. An all-composite fin is likely to be much lighter.

- The factor C_{sys} allows for on-board power systems such as the environmental control system, hydraulic generation and distribution, electrical generation, distribution, and flight controls. The use of this term indicates that the system weight depends on the cabin capacity as well as the MTOW. Design studies from the past suggest that $C_{sys} = 0.11$ may be considered as a realistic guess for a supersonic airliner.
- The term W_{fix} represents a nominal weight of items present in all passenger aircraft, independent of their size and flight speed: flight deck crew and its accommodation and instrumentation. From statistical information it is concluded that this component depends mainly on the cabin cross section. Calibration of the method for subsonic airliners indicated that for single-deck jetliners $W_{fix} \approx 500$ kg. This weight component forms a small fraction of the OEW for large airliners but a significant component for and executive aircraft, whereas according to [10] the complex variable geometry nose has no given weight problem.

Calibration of the factors of proportionality in Equation (3.2) is hampered by the fact that the available SCT information database is very small. Only the Concorde and the TU-144 have actually flown, whereas most of the information available on the SST projects developed in the USA during the 1960s has not been validated by prototype product information and flight performance. Indeed, the scarce data available on SCT development during the 1960s has become obsolete and needs to be updated in order to become useful for a future design.

It is worth noting that an expression similar to Equation (3.2) was discussed by Küchemann in [1] where he introduced "structural factors" with similar definitions and numerical values as C_{pay} and C_{to}. He used the following values available for Concorde: $C_{pay} = 1.50$ and $C_{to} = 0.35$. However, the definitions of these terms are not unambiguous and Küchemann commented that the structural factors are rather conservative values based on present technology and we should be able to achieve $C_{pay} = 1.0$ and $C_{to} = 0.25$ in a decade or two. This information is in accordance with the present text.

3.3 Fuel Weight

The total amount of fuel when taking off consists of mission fuel weight W_{misf} – derived from the design mission range R_d – and reserve fuel with weight W_{resf}. Mission fuel analysis of a synthesized design with known performance characteristics requires computation of the fuel burned during all sectors of the flight: take-off, climb, and acceleration to cruise altitude, cruising flight, descent, approach, and landing. Such an analysis is not feasible in the early design stage since most of the required information on the aerodynamic properties and the power plant is not yet available. An efficient solution of this problem is obtained by computing the required cruise fuel as accurately as possible and add a quasi-analytical allowance for additional fuel required for the other flight phases. This approach, published in [1], has been applied in the present case, albeit adapted to an SCV.

3.3.1 Mission Fuel

Since the conditions for quasi-stationary cruising flight are valid for subsonic as well as supersonic flight, Bréguet's generalized range equation can be used for computing the range in cruising flight (index cr):

$$R = R_H \mathcal{P}_{cr} \ln W_i / W_e \quad \text{with} \quad \mathcal{P}_{cr} \overset{\text{def}}{=} (\eta_0 L/D)_{cr} \quad \text{and} \quad R_H \overset{\text{def}}{=} H/g \approx 4{,}365 \text{ km.}$$
(3.3)

Equation (3.3) applies to a flight with constant Mach number and angle of attack. The calorific value of fuel per unit of mass H is incorporated in Equation (3.3) in the form of a reference range R_H for kerosene fuel and the aircraft gross weight for the initial and end conditions of the flight are W_i and W_e, respectively. The range parameter $\mathcal{P}_{cr}$ is identical to the flight efficiency $\eta_0 C_L / C_D$ which was introduced in Chapter 2 and prediction of the aerodynamic efficiency L/D will be treated in Chapter 7. Although Bréguet's equation represents the theoretical condition for the maximum range in quasi-steady flight, long-range flights are mostly executed with a stepped cruise/climb schedule.

According to [4] the fuel fraction required for such a sub-optimum flight can be approximated as follows:

$$\frac{W_{crf}}{W_i} = \frac{R_{cr}}{\mathcal{P}_{cr} R_H + 0.5 R_{cr}}.$$
(3.4)

It is worth noting that the factor 0.5 in the denominator of Equation (3.4) stems from a Fourier series approximation of the natural logarithm, indicating that the average specific range is assumed to be equal to the nominal specific range when 50% of the cruise fuel has been consumed. Since the cruising flight is treated as a quasi-steady motion, $\mathcal{P}_{cr}$ is considered to be constant and equal to its initial value. For a typical HSCT cruise fuel fraction $W_{crf}/W_i = 0.40$, the Bréguet range according to Equation (3.3) is 2% longer than the range obtained from Equation (3.4).

The reduced specific range during non-cruising flight sectors is accounted for by adding lost fuel, which is primarily determined by the plane's energy height at the top of the climb and by the reduced specific range at low flight speeds and altitudes. The lost fuel and the additional fuel required for maneuvering are taken into account by increasing the range in cruising flight by the lost range R_{lost}. The fuel required to fly the design mission is thus obtained by replacing the cruise range in Equation (3.4) by the equivalent all-out range $R_{eq} = R_{des} + R_{lost}$, resulting in

$$\frac{W_{misf}}{W_{to}} = \frac{R_{eq}}{\mathcal{P}_{cr}R_H + 0.5R_{eq}}. \tag{3.5}$$

The difference between cruising and off-design flight conditions can have a significant effect on the equivalent all-out range, in particular for short flights. An approximation for the lost range derived in [9] leads to the suggestion to use $R_{eq} \approx R_{des} + 0.20R_H$ for a long-range SCT mission.

3.3.2 Reserve Fuel

Accurate computation of the reserve fuel weight can be as cumbersome as computing the mission fuel weight. Fortunately, the early design stage of a subsonic airliner does not require such an effort since statistics indicate that reserve fuel weight is mostly between 4% and 5% of the MTOW and it is suggested to estimate the reserve fuel from

$$W_{resf} = C_{resf}W_{to}. \tag{3.6}$$

The subsonic flight conditions of SCT are, however, quite different from those during supersonic cruising and reduced flight efficiency values must be used for reserve fuel prediction. For Concorde this resulted in a reserve fuel weight of 6.5% of the mission fuel, not much different from its maximum payload fraction. This subject is associated with the discrepancy between cruise conditions and off-design flight conditions of an SCT and it is likely that an optimized supersonic configuration featuring variable geometry will not suffer from Concorde's exceptional reserve fuel fraction.

3.4 Take-off Weight and the Weight Growth Factor

The MTOW required to comply with the design mission is obtained by adding the maximum payload, the OEW according to Equation (3.2), the mission fuel according to Equation (3.5), and the reserve fuel according to Equation (3.6), resulting in the following closed form expression:

$$W_{to} = \frac{(1 + C_{sys}C_{pay})W_{pay} + W_{fix}}{1 - (\mathcal{P}_{cr}R_H/R_{eq} + 0.5)^{-1} - \{C_{sys}C_{af} + C_{pp}(T/W)_{to} + C_{resf}\}}. \tag{3.7}$$

Statistics have shown that, for a given fuel weight fraction, the empty weight fractions of subsonic and supersonic long-range aircraft are not very different. In the early phase of the design effort the designer's primary concern should therefore be focused on reducing the mission fuel weight required to comply with range requirements which is determined to a large extent by the range factor P_{cr}. Nevertheless, if during the downstream design process the empty weight of SCT has to be increased by a small amount ΔW_{OE}, this causes the gross weight to increase by ΔW_{TO} which is considerably larger than the initial ΔW_{OE}, provided the requirement is incurred that the design payload versus range performance must not deteriorate. This may result in time-consuming weight iterations, unless Equation (3.7) is used to solve for ΔW_{TO}. The result is expressed in terms of the "weight growth factor" defined as

$$\frac{\Delta W_{TO}}{\Delta W_{OE}} = \frac{1}{1 - (P_{cr}R_H/R_{eq} + 0.5) - C_{sys}C_{af} - C_{resf}}, \tag{3.8}$$

where the mission fuel weight fraction is the most influential term that increases nearly proportional to the equivalent design range. Since for a long-range SCV the total fuel load may exceed 50% of the MTOW, the growth factor increases rapidly with the range and tends asymptotically to infinity for the ultimate (equivalent) range,

$$(R_{eq})_{ult} = \frac{P_{cr}R_H}{1 - C_{sys}(C_{af} - C_{resf})^{-1} - 0.5}. \tag{3.9}$$

In order to avoid redesign iterations after an empty weight increment from becoming a diverging process, the range according to Equation (3.9) must be (considerably) longer than the required design mission range.

It is worth noting that the present method for predicting the weight breakdown is basically valid for subsonic as well as supersonic cruise vehicles. However, if a subsonic and a supersonic transport designed for the same payload and range are compared with the same weight prediction method, their OWE, MZFW, and MTOW exhibit large differences. The dominating reason for this is found mainly in the difference between the range factors P_{cr} of the two designs. For example, the range factor for a present day long-range subsonic airliner in cruising flight amounts to $P_{cr} \approx 8.0$ typically, whereas design studies of a second generation supersonic airliner indicate that according to Figure 2.4 $P_{cr} \approx 5.0$ may be achievable. The next section will illustrate how the methodology exposed in this chapter can be applied to make an early assessment of the design weight sensitivity to variations in mission requirements and technological input parameters.

3.5 Example of an Early Weight Prediction

The following example published in [11] is based on design specifications of the CISAP Joint Research and Technology Cooperation agreement between Airbus

Industrie and AREA partners DLR, NLR, ONERA, and QinetiQ. The baseline mission is to carry 250 passengers in mixed-class seating over a distance of 10,000 km at Mach 2.0 cruise speed. However, the CISAP project participants studied alternative designs with cruise Mach numbers 1.30, 1.60, and Mach 2.0. The following example is based on mission requirements for the baseline ESCT configuration described in Section 1.5, summarized as follows:

- Payload: 250 passengers in three class layout.
- Supersonic cruise speed: Mach 2.0. A cruise/climb profile can be assumed with the aircraft flying at the optimum altitude for each weight. Subsonic cruise speed: Mach 0.95.
- Design mission range: 10,000 km, including an outbound subsonic leg of 50 km, flown with the same specific range as the supersonic cruise sector.
- Reserves: 4% block fuel; 463 km diversion at 10,500 m altitude; hold for 30 min. with 463 km h^{-1} at 4,500 m altitude.
- Take-off distance not exceeding 3000 m.
- Landing approach speed not exceeding 300 km h^{-1}.

3.5.1 MTOW Sensitivity

The basic weight weight distribution is derived by inserting the following educated guesses into Equation (3.7):

- Fuselage group weight: 150% of the volumetric payload; hence, $C_{pay} = 1.5$.
- The wing structure weight is sensitive to the parameters affecting the aerodynamic efficiency, in particular the MTOW and the wing planform geometry. Since wing geometry is the most influential component for optimizing the vehicle's configuration, its structure weight fraction is treated as a selection variable.
- The power plant weight is based on a typical take-off thrust/weight ratio of 0.35, an engine dry weight/thrust ratio of 0.20 and an additional 30% for engine accessories, intakes and engine mounts, resulting in an installed weight fraction of 0.091.
- The undercarriage weight fraction of 0.044 is assumed for a three-leg main gear.
- The reserve fuel fraction $C_{resf} = 0.055$ is midway between subsonic airliners and Concorde.
- The on-board power systems weight equals 12% of the payload-dependent as well as the MTOW-dependent OEW components. Hence, $C_{sys} = 1.12$.

These assumptions yield a structural factor $C_{af} = 0.135 + W_{wing}/W_{to}$ and, for a lost range $R_{lost} = 0.20R_H$, the equivalent all-out range amounts to 10,870 km. The zero-fuel weight (ZFW) is obtained from summation of the previous items,

$$W_{zf} = 75,540 + 0.206W_{to} + 1.12W_w \tag{3.10}$$

and the take-off gross weight required to comply with the above mentioned requirements is found by substitution of the structural factors into Equation (3.7),

$$W_{to} = \frac{75,540}{0.794 - [1.12(W_{wing}/W_{to}) + (P_{cr}R_H/R_{eq} + 0.5)]^{-1}} \quad \text{(kg)}. \qquad (3.11)$$

As an example of a conservative design we assume a "year 2000 feasible" aerodynamic efficiency $L/D = 9.50$ and an overall propulsive efficiency $\eta_0 = 0.42$ which yields a range parameter $P_{cr} = 4$. For an assumed wing structure weight fraction of 0.10, the MTOW required to realize the mission equals 365,000 kg. The corresponding payload fraction is about 15% higher than Concorde's payload fraction. This performance must be considered as unsatisfactory for next-generation SCT and it is necessary to investigate technological advancements required to increase the range parameter and/or design models leading to reduced wing structure weight. Figure 3.1 depicts the MTOW as a function of the wing structure weight fraction and the range parameter in cruising flight P_{cr}. It shows that, due to the high fuel weight fraction, the MTOW is more sensitive to the range parameter than to the wing weight fraction – a typical property of long-range aircraft designed for supersonic cruising. In this initial weight study we therefore focus on improving the range parameter. For instance, a realistic aim for advanced future second generation SCT could be the achievement of $P_{cr} = 4.0$ according to Figure 2.4; that is, 33% higher than Concorde's range parameter. For a wing structure weight fraction of 0.10 the MTOW predicted according to Equation (3.11) amounts to 304,000 kg, corresponding to an OWE fraction 0.42 and a total fuel weight fraction of 0.49. The resulting payload fraction of 0.09 is one third higher than Concorde's payload fraction, which was achieved with $P_{cr} \approx 3.0$. Achieving $P_{cr} = 4.50$ will,

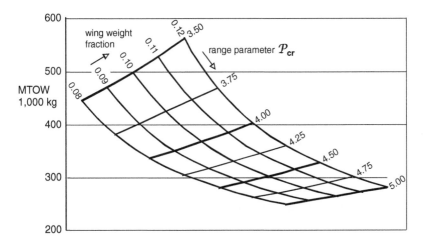

Figure 3.1 MTOW of Mach 2.0 SCT versus the range parameter and the wing weight.

however, require a major effort from the aerodynamic designers involved in the development of an advanced second generation CST with similar stringent top level requirements as assumed for this example.

The weight sensitivity of the design cycle is further clarified by assuming that downstream in the more detailed design process the wing structure weight appears to be 10% heavier then initially predicted. In that case the MTOW increases by 5.5% to 385,000 kg, whereas in the case of a 10% reduction of the range parameter the MTOW would increase by 20% to 440,000 kg. If, on the other hand, the aerodynamic and the propulsive efficiencies could both increase by 10%, the MTOW would decrease by a sensational 24% to 277,000 kg. The message is that fairly modest future improvements in the structural, aerodynamic, and propulsion characteristics compared to values which were considered realistic in the year 2000 may lead to a substantial weight reduction for a next-generation advanced SCT design.

3.6 Productivity and Energy Efficiency

The payload fraction that can be carried for a given range is derived from Equation (3.7),

$$\frac{W_{\text{pay}}}{W_{\text{to}}} = \frac{1 - \{C_{\text{sys}}C_{\text{af}} + (\mathcal{P}_{\text{cr}}R_{\text{H}}/R_{\text{eq}} + 0.5)^{-1} + C_{\text{resf}}\}}{1 + C_{\text{sys}}C_{\text{pay}} + W_{\text{fix}}/W_{\text{pay}}}. \tag{3.12}$$

Equation (3.12) is useful to compute the range for the "conservative design" achievable with the same MTOW and a mixed-class seating for 250 passengers, which is more representative of the operation of a long-range SCT. This results in an achievable distance of 11,000 km, an attractive range for this comfortable high-speed transport over long routes. Maximizing the payload fraction has an impact on the productivity, the environmental quality and the operating costs. Although top level requirements normally specify constraints on the seating capacity as well as on the design range, it is useful to have an insight into the effects of varying both characteristics simultaneously. The following considerations inspired by [5] emphasize the importance of maximizing the productivity and the energy efficiency of a transport aircraft. The payload versus range diagram in Figure 3.2 illustrates the transport capability of the airplane and its profit potential. The upper and right portion of the chart represents the envelope of payload and range capability consisting of three straight sectors. The horizontal sector of the diagram represents the maximum weight – or volume-limited payload – its length equals the harmonic range when taking off with the maximum MTOW. The oblique sector is the range achievable when the plane takes off with the MTOW and reduced payload. The near-vertical sector

Figure 3.2 Definition of the profit potential [5, 6].

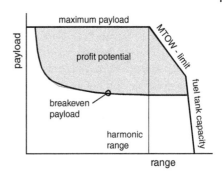

defines the range achievable with fuel tanks filled to their capacity. The area under the payload-range envelope represents the productivity, an important feature of a transport aircraft for judging its commercial value. The productivity is primarily determined by the product of the maximum payload and the harmonic range.

A high productivity forms an important factor to increase the potential operational profit. On the other hand, the operating costs incurred with achieving a high productivity increase with increasing MTOW. We intend to demonstrate that, for a given MTOW, there exists a stretch of design mission ranges for which the productivity has a (near-)maximum value. The curved envelope in Figure 3.2 on the lower and left region of the payload-range chart defines the payload or the number of seats passengers that must be carried to pay for the airplane's operation. The curved envelope on the lower and left region of the payload-range chart defines the payload and the number of passengers that must be carried to pay for the airplanes operation For very short ranges all seats installed in the cabin must be filled to break even, which is ascribed mainly to the fixed cost per flight and the dominance of the non-cruising flight sectors with a low average costs, so that the airplane costs per unit of distance flown increases progressively when the trip gets shorter. The maximum payload times range capacity minus the breakeven number of passengers is referred to as the profit potential. The ratio between the profit potential and the productivity can be considered as the airliner's economic efficiency.

3.6.1 Range for Maximum Productivity

The achievable productivity for a specified MTOW is proportional to the product of the (maximum) payload fraction and the harmonic range R_{des}. Referred to as the productivity parameter this figure of economic merit amounts to

$$P_{prod} \stackrel{\text{def}}{=} \frac{W_{pay}}{W_{to}} \frac{R_{eq}}{R_H}. \tag{3.13}$$

The dimensionless productivity is computed for various values of the range factor P_{cr} according to Equation (3.2) for the example second generation SCT treated in Section 3.5. The result depicted on Figure 3.3 is explained by the fact that for short ranges the increasing range is dominating, whereas for long ranges the decreasing payload fraction dominates. Since the curves exhibit a shallow optimum the conclusion can be drawn that, from the point of view of conceiving an airliner with flexible operational capabilities, the preferred design range should be longer rather than shorter than this optimum. The maximum P_{prod} is achieved for $R_{des}/R_H \approx 0.45 P_{cr}$. The "advanced future design" introduced previously, for which $P_{cr} = 4.4$, is sized for a range of 10,000 km, thereby accurately satisfying this condition for maximum productivity. It is interesting that the Concorde, for which $P_{cr} \approx 3$, the maximum productivity is achieved for a range of 6500 km, which is actually Concorde's maximum range. These results may be compared with present-day long-range airliners having a typical range parameter $P_{cr} = 8$, for which the condition for maximum productivity leads to a theoretical optimum design range according to Figure 3.3 of 16,000 km.

3.6.2 Energy Efficiency

The fuel energy efficiency of a passenger transport is traditionally defined as the seat-kilometer production per unit of fuel volume consumed during a given distance traveled,

$$\text{FEE} = \frac{\text{number of seats}}{\times} \text{distance traveled volume of fuel consumed.}$$

$$(3.14)$$

A similar characteristic is the payload fuel efficiency, specifying the payload-range productivity per unit of fuel weight consumed,

$$\text{PFE} = \frac{\text{payload} \times \text{distance traveled}}{\text{weight of fuel consumed}}.$$

$$(3.15)$$

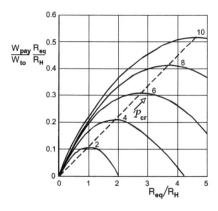

Figure 3.3 Definition of the productivity [5, 6].

Applied to a given period of commercial operations, the total seat-kilometer or payload-kilometer produced represents the income potential, whereas the total amount of fuel consumed forms a large proportion of the operational expenses. Interpreted as a basic cruising flight performance, the energy efficiency can be defined alternatively as the ratio of transport work produced and fuel energy required to deliver this work. Applied to the instantaneous cruise performance, the momentary energy efficiency is proportional to the specific range V/F:

$$E_{en} \overset{\text{def}}{=} \frac{W_{pay}\Delta R}{\Delta Q_f \rho_f H} = \frac{W_{pay}}{R_H}\frac{V}{F}, \tag{3.16}$$

where ρ_f denotes the specific mass of fuel and $\Delta R/\Delta Q_f$ is the distance flown per amount of fuel volume consumed. Equation 3.16 defines a dimensionless energy efficiency and, hence, one might be tempted to compare this with the usual definition of a mechanical or thermodynamic efficiency. This would, however, be an erroneous interpretation of the term "transportation work" since the energy level of the payload is not changing during (quasi-)stationary flight. Combination of Equations (3.16) and (3.17) yields the FEE rewritten in terms of the range parameter and the payload fraction,

$$\text{FEE} \overset{\text{def}}{=} \frac{N_{seat}\Delta R}{\Delta Q_f} = E_{ref}P_{cr}\frac{W_{pay}}{W_g} \quad \text{with} \quad E_{ref} \overset{\text{def}}{=} \rho_f H \frac{N_{seat}}{W_{pay}}, \tag{3.17}$$

where W_g denotes the aircraft momentary gross weight. For a typical payload weight per seat of 95 kg and a fuel energy density $H = 34.5$ MJ L^{-1}, we find $E_{ref} = 37$ seat-kilometer per liter.

A more adequate criterion is the fuel energy efficiency referred to the mission fuel burnt per seat in relation to the range $\text{FEE} = N_{seat}R_{mis}\rho_f g/W_{misf}$. Applying this criterion to hypothetical "advanced future SCT" with $P_{cr} = 4.4$ and a mid-cruise payload fraction of 0.115, and combining it with the computed mission fuel of 134,000 kg according to Section 3.3, we find an FEE of 18 seat-kilometers per liter. Equation (3.17) results in an average 17.6 seat-kilometers per liter fuel for the complete mission – 3% below the instantaneous FEE in mid-cruising flight. Although this compares favorably with Concorde's (estimated) FEE of 7.3 seat-kilometers per liter it is worth noting that a modern subsonic jetliner with the same payload-range productivity produces a typical 50 seat-kilometers per liter. This makes it clear that the SCT, in spite of its much higher block speed, will have a substantial problem in being competitive on the commercial market.

3.6.3 Conclusion

The early weight prediction method exposed in this chapter and its application to second generation HSCT suggest that, compared to Concorde's performances, the payload-range production can be improved considerably, provided the flight efficiency in high-speed flight can be increased by at least 50%, making future advanced SCT carrying 250 passengers over the 10,000 km range at Mach 2.0 an attractive alternative for flying trans-Pacific routes. Compared to present-day subsonic airliners with similar payload-range production, such an airliner would offer a block time reduction of up to 50%, leading to a productivity increase of 100% with 2.4 times Concorde's FEE. Since this figure of merit is (at least to a considerable extent) under the control of the advanced design team, the FEE may be considered as a prime candidate to be used as a figure of merit for optimizing the design.

Bibliography

1 Küchemann, D. *The Aerodynamic Design of Aircraft*. 1st ed. Oxford: Pergamon Press; 1978.

2 Fielding, J.P. *Introduction to Aircraft Design*. Cambridge Aerospace Series 11, Cambridge: Cambridge University Press; 1999.

3 Whitford, R. *Evolution of the Airliner*. Ramsbury: The Crowood Press; 2007.

4 Torenbeek, E. *Advanced Aircraft Design – Conceptual Design, Analysis and Optimization of Subsonic Civil Airplanes*. Chichester: John Wiley and Sons Ltd; 2013.

5 Peyton Autry, C., and Baumgaertner P.J. The Design Importance of Airplane Mile Costs Versus Seat Mile Costs. SAE Paper No. 660277, April 1966.

6 Engineering Sciences Data Unit. Lost Range, Fuel and Time to Climb and Descent: Aircraft with Turbojet and Turbofan Engines. ESDU Data Sheet No. 74018, August 1974. Available at: https://www.esdu.com/cgi-bin/ps.pl? sess=unlicensed_1191114135640rsg&t=doc&p=esdu_74018a

7 Engineering Sciences Data Unit. Approximate Methods for Estimation of Cruise Range and Endurance: Aeroplanes with Turbojet and Turbofan Engines. ESDU Data Sheet No. 73019, revised May 1982. Available at: https://www.esdu.com/cgi-bin/ps.pl?sess=unlicensed_1191114135746pvj& t=doc&p=esdu_73019c

8 Bos, A.H.W. Multidisciplinary Design Optimization of a Second-Generation Supersonic Transport using a Hybrid Genetic/Gradient-Guided Algorithm. Doctoral Thesis TU Delft; 1996.

9 Torenbeek, E. Cruise Performance and Range Prediction Reconsidered. *Progr. Aerospace Sci.* 33(5/6):285–321; 1997.

10 Collard, D. Concorde Airframe Design and Development. Von Kármán Institute for Fluid Dynamics, J.J. Ginoux Lecture, 17 November 2000. Available at: https://www.jstor.org/stable/pdf/44548119.pdf?seq=1#page_scan_tab_contents

11 Torenbeek, E., Jesse E., and Laban M. Conceptual Design and Analysis of a Mach 1.6 European Supersonic Commercial Transport Aircraft. CISAP Deliverable D2-WP1.2, National Aerospace Laboratory NLR, NLR-CR-2003-384, 2003.

4

Aerodynamic Phenomena in Supersonic Flow

Compressibility effects were recognized and treated fundamentally by E. Mach (1838–1916), L. Prandtl (1875–1953), T. Von Kármán (1881–1963), J. Ackeret (1898–1981), and several other scientists. The development of supersonic aerodynamic theory initially advanced concurrently with low-speed aerodynamic theory since propeller aircraft that could reach speeds of 900 km h^{-1} in steep diving flight experienced serious stability and control problems due to the compressibility of air. However, the development of applications to high-subsonic jet airliners and supersonic vehicle design technology rapidly has sped up since the 1930s, when the progress of gas turbine engine technology made it clear that supersonic flight would soon become a reality.

4.1 Compressibility of Atmospheric Air

Although the first turbojet-powered aircraft had enough thrust to allow supersonic flight, they were also subject to adverse flight dynamics behavior since the aforementioned aerodynamic phenomena had a profound influence on the forces and moments acting on the plane. In the present world of large-scale applications of high-speed aircraft, designers involved in the development of a supersonic cruise vehicle (SCV) should have a basic understanding of aerodynamic phenomena in high-speed flight, such as shock and expansion waves. Moreover, the prediction of aerodynamic properties such as the airplane's aerodynamic efficiency and the variation of its aerodynamic center with Mach number are essential elements during the early development of a supersonic transport aircraft configuration. An SCV must be able to cruise efficiently at supersonic as well as high-subsonic Mach numbers similar to those of high-subsonic airliners. Moreover, supersonic transport aircraft must have good aerodynamic properties in low-speed flight for taking off and landing and, since the effects of compressibility have a major effect on aerodynamic phenomena observed at the complete range of operational Mach numbers,

Essentials of Supersonic Commercial Aircraft Conceptual Design, First Edition. Egbert Torenbeek.
© 2020 Egbert Torenbeek. Published 2020 by John Wiley & Sons Ltd.

the conceptual designer should have good insight into the aerodynamic phenomena affecting the plane's behavior.

The aerodynamic phenomena to be discussed in this book will be limited to those occurring at speeds below Mach 5.0, which implies that the air can be treated as a calorific perfect gas with constant values of the specific heat. Since the airplane's geometry is closely related to its most essential aerodynamic properties, some selected elements of classical linearized solutions to theoretical models are presented. The present chapter offers an abstract of topics treated in [8] on high speed flow phenomena around body shapes representative of aircraft components as well as applications of the theory. More in-depth treatments of aerodynamic phenomena around high-speed flight vehicles to which attention must be paid in the aerodynamic design phase can be found in textbooks such as [4, 9], and [10].

4.1.1 Speed of Sound and Mach Number

An infinitesimal pressure disturbance such as a sound wave is transmitted in the atmosphere at the sonic (or acoustic) velocity. The propagation of sound is closely related to the transfer of momentum between colliding molecules, which depends on their average speed, whereas the average kinetic energy of molecules is proportional to the (local) temperature of the medium. The implication is that, according to the kinetic theory, the molecules of a gas are moving with an average velocity of $\sqrt{8RT/\pi}$, where R and T denote the gas constant and the temperature, respectively. It has been observed that the sonic velocity is about 75% of this value. A derivation based on the conservation equations of continuity, momentum and energy for isentropic flow through a stationary sound wave in a moving gas yields the following expression for the speed of sound:

$$a = \sqrt{dp/d\rho} = \sqrt{\gamma p/\rho} = \sqrt{\gamma RT}, \tag{4.1}$$

where ρ denotes the air density and γ the ratio of specific heat at constant pressure c_p and volume c_v. For atmospheric air, the gas constant amounts to $R = 287$ J kg^{-1} K^{-1} and the ratio of specific heat amounts to $\gamma = 1.40$. Hence, it follows that the sonic velocity equals $a \approx 20\sqrt{T}$ m s^{-1} and varies between 340 m s^{-1} at sea level and 295 m s^{-1} in the stratosphere. The most convenient index to assess whether the flow can be considered as incompressible is the Mach number $M \stackrel{\text{def}}{=} V/a$, defined as the ratio of the (local) flow velocity to the local sonic velocity.

Present-day high-subsonic airliners are designed to travel at cruise Mach numbers typically between $M = 0.7$ and $M = 0.9$. In this flight regime aerodynamic effects such as shock waves, shock-induced flow separation, and buffeting may occur associated with the compressibility of air. Since extensive occurrence of these phenomena deteriorates aerodynamic performance and flying qualities, operational constraints are imposed on the vehicle's flight envelope to avoid objectionable or even dangerous conditions. Designers of modern jetliners are familiar with the principles of transonic aerodynamics and they are aware of undesirable aerodynamic phenomena that may occur in high-speed flows and how to conceive geometries by which they can be avoided.

4.1.2 Compressible and Incompressible Flows

A flow in which the density is independent of the pressure is regarded as incompressible. Since very little pressure is needed to change the volume of a certain quantity of atmospheric air, its compressibility is several orders of magnitude greater than that of liquids, which are hardly compressible. Consequently, the often made assumption that low-speed airflow can be considered to be incompressible may need explanation.

The amount by which air can be compressed is given by its compressibility, defined as $\tau \overset{\text{def}}{=} (dv/dp)/v$. This quantity can be physically interpreted as the fractional increase in volume dv of an element of air per unit change in pressure p exerted on its external surface. Substitution of the specific volume $v = 1/\rho$ and the relationship between the compressibility and the sonic velocity leads to

$$\tau = (\rho a^2)^{-1} = (\gamma p)^{-1}. \tag{4.2}$$

The formal interpretation of this result is that, for the assumption that the flow is incompressible to be true, incompressible flows are theoretically zero Mach number flows. Although this supposition conflicts with the physical reality, it is often accepted since experience has taught that at low-subsonic flight speeds the density variation of the flow surrounding the aircraft plays a sub-dominant role in most aerodynamic phenomena. Assuming the flow to be incompressible appears to be very reasonable for solving many low-speed aerodynamic problems and is widely accepted for flight speeds below $M = 0.3$. For higher subsonic flight speeds, corrections must be made for compressibility effects.

4.2 Streamlines and Mach Waves

Figure 4.1(a) depicts a blunt-nosed airfoil section in two-dimensional subsonic flow. The streamlines, representing the path of the airflow particles, are curved in a large region around the airfoil to give way to the approaching object. In this situation a metaphor can be used stating that "the upstream air is prepared for the approaching body and gives way gradually". This results in smoothly curved streamlines in front of and alongside the body, whereas their curvature disappears at long distances away from the airfoil.

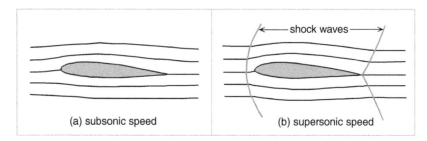

Figure 4.1 Streamlines and shock waves past a stationary airfoil in two-dimensional air flows moving at low and high speeds.

The compressibility of the air at high speeds has far-reaching consequences for the aerodynamic phenomena in the flow past the same airfoil. Figure 4.1(b) illustrates that at supersonic speeds the air particles in the flow upstream of the airfoil follow straight trajectories since they remain "unaware of the approaching object". At some distance ahead of the airfoil nose a bow shock wave is formed where the air pressure increases abruptly and the particle speed is reduced to subsonic velocity, whereas oblique shock waves are observed above and below the airfoil trailing edge. The streamlines behind the airfoil are nearly straight.

4.2.1 Sound Waves

Basic differences between subsonic and supersonic flow properties can be explained by looking at the propagation of waves due to sound pulses emitted by a point source that moves through a static atmosphere at different speeds. This point source could be, for instance, a tone generator periodically producing infinitesimally weak pressure pulses that are emitted as sound waves propagating in all directions. Figure 4.2 shows how sound waves

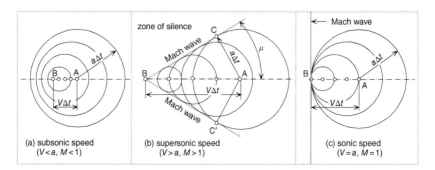

Figure 4.2 Propagation of sound waves emitted by a moving point source in a static atmosphere and generation of Mach waves.

propagate in a two-dimensional plane dependent on the velocity at which the source moves.

(a) A source moving at subsonic speed emits waves at different points taking the form of circular fronts. While the source travels along a straight line from A to B during a time interval Δt covering a distance AB $= V\Delta t$, pressure pulses have been produced depicted as circles with radius decreasing from $a\Delta t$ to zero. The source itself stays within the collection of pressure disturbances represented by the circular fronts. A similar reasoning applies to a body moving at subsonic Mach number through stationary air.

(b) The situation is markedly different when the speed of the point source is increased beyond the sonic speed, since it moves with supersonic speed during the same time interval Δt as in the previous example. The supersonic source moves faster than the emitted sound waves, it remains outside the circular fronts of the produced perturbations, and the distance AB $= V\Delta t$ is longer than the radius $a\Delta t$. The zone in which the emitted sound waves have stretched out is limited by two infinitesimally weak Mach waves that are tangent to the circular disturbance fronts. Figure 4.2(b) illustrates the Mach angle μ between these oblique waves and the trajectory of the point source, defined by

$$\sin \mu = \frac{a\Delta t}{V\Delta t} = \frac{a}{V} = \frac{1}{M} \quad \rightarrow \quad \mu = \sin^{-1}\frac{1}{M}. \tag{4.3}$$

The waves emitted by a sound source traveling along a straight line in a 3D space have a conical envelope of Mach waves with an apex angle at point B equal to 2μ. Plane wave envelopes may be formed by a line source such as the straight and sharp leading edge of a wing, where the Mach waves generate two oblique flat planes with an angle 2μ between them.

(c) When the point source travels at the sonic velocity $V = a$, the disturbances are not propagated ahead of the source but only behind it with relative velocity $2a$. This results in a collection of circular fronts with increasing diameters proportional to the time δt during which the point source has been traveling away from point A. The disturbances then build up in the source into a plain Mach wave perpendicular to the flow dividing the region which is affected from that which is not. In the three-dimensional space this is a planar Mach wave normal to the trajectory AB that spreads out to infinity in all directions.

It can be stated that air particles in front of a supersonic body travel along straight paths since they are "unaware of the approaching body and do not give way to it". The region in front of a shock wave attached to the nose of a supersonic object is therefore known as a zone of silence in which the air particles travel along straight streamlines. Since perturbations emitted by a supersonic point source or flow disturbances caused by the supersonic object stay behind the conical Mach wave and the nose shock wave, respectively, these regions are sometimes called zones of action.

4.3 Shock Waves

The flow field around a supersonic flight vehicle features shock waves – phenomena which are to some extent similar to the Mach waves described in Section 4.2.1. A shock wave is a non-isentropic phenomenon since it is much stronger than the sound waves emitted by a point source. The formation of a shock wave occurs when the flow decelerates in response to a sharp increase in pressure or when the flow encounters a sudden compressive change in a direction invoked by a high-speed airplane component. The strength of a shock wave is determined primarily by the body's geometry, in particular the slope of the local surface relative to the flight path. For example, a slender body of revolution with a sharply pointed nose causes a relatively weak near-conical shock wave emanating from the nose, whereas a two-dimensional blunt-nosed body will generate a strong bow shock wave in front of it. Large pressure perturbations are also formed at locations where the body surface exhibits a pronounced kink or a discontinuous variation of the cross-sectional area. In general, any discontinuity in the body cross sectional area distribution normal to the flow will lead to pressure disturbances causing phenomena such as shock waves not observed in sub-critical flow.

The term shock wave defines the extremely thin layer of air in which the state properties change in a distance related to the mean-free path length of the molecules, which amounts to approximately 3.7×10^{-3} mm at standard sea level conditions. The thickness of a shock is typically between three and five times the mean free-path length; that is, between one and two times 10^{-2} mm[1]. In going through the shock wave, the density and pressure of the flow are increased, but the velocity is reduced. This process is associated with energy dissipation and increased entropy, but mathematically a shock wave can be treated as a discontinuity. The viscosity of the air inside the shock wave converts kinetic flow energy into heat, and the associated entropy increment causes a stagnation pressure loss of the flow downstream of a body, which becomes manifest as wave drag.

If a shock wave interacts with a boundary layer it promotes the local flow to separate, usually resulting in another source of drag. If, however, a pressure rise due to a shock wave acts on the lower wing surface, this can be utilized as a contribution

---

1 Shocks waves are usually depicted as a double line to distinguish them from streamlines.

to the lift. Altogether, shock waves may cause a significant drag penalty, loss of lift and a reduced aerodynamic efficiency. The shock waves generated by a supersonic airplane travel through the atmosphere over long distances outward and downward behind the plane and, when arriving on the earth's surface, they produce a sonic boom. Shock waves cannot be avoided altogether but it is of utmost importance to minimize their strength during critical phases of the flight.

4.4 Normal Shock Waves

Figure 4.3(a) depicts a channel with constant cross sectional area in which a uniform supersonic flow enters the channel with velocity V_1, Mach number M_1, density ρ_1 and pressure p_1. Normal to the oncoming flow, air particles decelerate abruptly to subsonic speed through a stationary shock wave forming a planar surface perpendicular to the streamlines of the oncoming flow. The particles continue their original direction and hence the streamlines are not kinked. For given conditions of the oncoming flow (index 1), the flow properties behind the shock wave (index 2) can be computed by combining the conservation laws of mass ($\rho_1 v_1 = \rho_2 v_2$), momentum ($p_1 + \rho_1 v_1^2 = p_2 + \rho_2 v_2^2$) and energy ($h_1 + v_1^2/2 = h_2 + v_2^2/2$). The enthalpy h is the sum of the internal heat and the kinetic energies of the medium. Since no mechanical or thermal energy is added, the flow through a shock wave is adiabatic, but not isentropic. Derivations in [2] and [9] of the relation between the shock properties in front of and behind

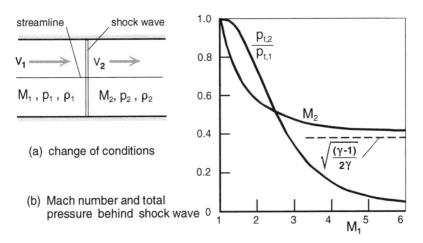

(a) change of conditions

(b) Mach number and total pressure behind shock wave

Figure 4.3 Stationary normal shock wave in uniform flow and properties of the downstream flow depending on the Mach number of the oncoming flow.

the shock wave assume the air inside the shock to ignore heat conduction, leading to the following result for the Mach number behind the shock:

$$M_2 = \sqrt{\frac{M_1^2 + 5}{7M_1^2 - 1}} \quad \text{for} \quad \gamma = 1.40. \tag{4.4}$$

This result states that the Mach number behind the wave is a function of the Mach number ahead of the wave only. Moreover, $M_2 = 1$ if $M_1 = 1$. In this case the shock is infinitely weak; hence it is a Mach wave. In real flow, only the solution $M_1 > 1$ and $M_2 < 1$ is possible. Consequently, a normal shock wave can exist only in a supersonic oncoming flow and the flow downstream of the shock is always subsonic. As a corollary, Equation (4.4) proves that the velocity at which a normal shock wave is propagated in static air is higher than the speed of sound and the ratios of the densities and pressures behind and in front of the shock are derived from the continuity equation,

$$\frac{\rho_2}{\rho_1} = \frac{(\gamma + 1)M_1^2}{(\gamma - 1)M_1^2 + 2} \quad \text{and} \quad \frac{p_2}{p_1} = 1 + \frac{2\gamma}{\gamma + 1}(M_1^2 - 1), \tag{4.5}$$

whereas the (considerable) temperature increase when passing the shock increases with the Mach number according to

$$\frac{T_2}{T_1} = \frac{p_2/p_1}{M_1^2}. \tag{4.6}$$

The density ratio increases rapidly to five at Mach 5.0 and approaches the value six for $M_1 \to \infty$. The relation between the pressures and densities through the shock is described by the Rankine Hugoniot equation,

$$\frac{p_2}{p_1} = \left[1 - \frac{\gamma - 1}{\gamma + 1}\frac{\rho_2}{\rho_1}\right]\left[\frac{\rho_2}{\rho_1} - \frac{\gamma + 1}{\gamma - 1}\right]^{-1}. \tag{4.7}$$

Equations (4.4) and (4.6) clearly show that the stronger the shock wave, the faster it travels through the atmosphere, whereas Figure 4.3(b) shows that M_2 decreases monotonically with M_1 and approaches the theoretical lower limit $M_2 = 0.378$ for $M_1 \to \infty$.

4.4.1 Effects of Normal Shock Waves

The abrupt pressure increase caused by a shock wave has a large influence on the general flow field around, and the pressure distribution on, the vehicle. The pressure, density, and temperature behind a normal shock wave are larger than in front of it and when passing through the shock wave the flow experiences a considerable entropy loss. Figure 4.3(b) clearly demonstrates the radical effect of increasing the oncoming flow velocity on the stagnation pressure and thereby explains the wave drag associated with a normal shock. Normal shocks occur frequently in channel flows such as the intake of a gas turbine engine and the total pressure loss from a normal shock wave has far-reaching consequences for the design of a supersonic engine installation. In particular, the total pressure loss due to a normal shock inside the air intake of a supersonic engine causes a significant efficiency loss and has far-reaching consequences for its mechanical design. Normal shock waves can also occur in transonic external flows around a lifting surface, where they terminate a local supersonic region on top of the airfoil. And the interaction between a strong shock wave and the boundary layer promotes separation of the flow, which is often the main cause of the steep drag rise at high subsonic speeds.

4.5 Planar Oblique Shock Waves

Depending on the geometry of the body generating it, a shock wave is mostly curved in three dimensions. Figure 4.4 depicts a two-dimensional flow along a concave wall that is compressed at a sharp corner, a geometry representative of an obstacle such as a wedge-shaped airfoil leading edge[2].

The upstream flow with speed V_1 and Mach number M_1 is forced to follow the turn angle δ of the wall towards the air stream. This results in an oblique shock wave occurring in the corner point, which increases the pressure of the the downstream flow to p_2 and decreases its Mach number to M_2. The imposed instantaneous flow deflection θ depends on the geometry of the body causing the shock wave and is manifest in kinked streamlines. In the present example of a planar deflected wall, the shock wave is also planar and, hence, straight in the two-dimensional plane (with $\theta = \delta$). However, strong oblique shocks are often curved, leading to variation of θ along the shock. Since the disturbance associated

Figure 4.4   Planar oblique shock in two-dimensional supersonic flow deflected by a concave surface with a sharp turn angle towards the flow.

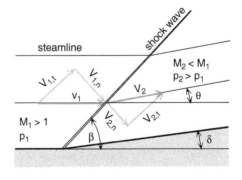

2 Analysis of a three-dimensional curved shock wave is outside the framework of the present text.

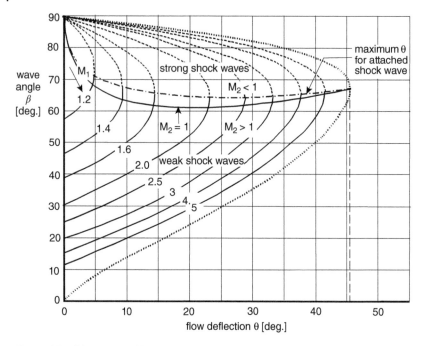

Figure 4.5 Diagram specifying the wave angle of a planar oblique shock as a function of the flow deflection and the oncoming flow Mach number [11].

with the shock wave is finite, its propagation speed is greater than the speed of sound and the shock wave inclination angle β towards the oncoming flow is greater than the Mach angle of the oncoming flow. The physical mechanism creating oblique shocks in a supersonic flow is essentially the same as that for Mach waves, described in Section 4.2.1. In fact, a Mach wave can be seen as an infinitely weak oblique shock.

In reality, an oblique shock causes changes in the flow similar to the normal shock, which can be conceived as a special category of the oblique shock: the case of $\beta = 90°$. The derivation of relations between the geometry creating an oblique shock wave and the wave properties is therefore similar to those of a normal shock described in Section 4.2.1, except that the direction of the upstream and downstream flows are different.

Evaluation of the shock geometry in Figure 4.5 leads to the following conclusions:

- There is no mechanism to increase or decrease the tangential component of the flow velocity across the shock, which dictates that $V_{1,t} = V_{2,t}$. This can be proven by combining the continuity and the momentum equations applying to the upstream and downstream flows.

- Application of the conservation equations for momentum and energy leads to the result that the changes across the oblique shock are governed only by the velocity components normal to the wave.

The normal shock equations leading to Equation (4.4) can be applied to those for the oblique shock by replacing M_1 by $M_1 \sin \beta$, the component of the upstream flow Mach number normal to the oblique shock wave. This yields the Mach number of the downstream flow

$$M_{2,\text{n}} = \sqrt{\frac{(\gamma - 1)(M_1 \sin \beta)^2 + 2}{2\gamma(M_1 \sin \beta)^2 - (\gamma - 1)}} \tag{4.8}$$

and the ratios of the pressure and density

$$\frac{p_2}{p_1} = 1 + \frac{2\gamma}{\gamma + 1}[(M_1 \sin \beta)^2 - 1] \quad \text{and} \quad \frac{\rho_2}{\rho_1} = \frac{(\gamma + 1)(M_1 \sin \beta)^2}{2 + (\gamma - 1)[M_1 \sin \beta]^2}. \tag{4.9}$$

The temperature ratio is obtained from $\frac{T_2}{T_1} = \frac{p_2/p_1}{\rho_2/\rho_1}$ and the downstream Mach number is derived from the normal shock geometry $M_2 = M_{2,\text{n}} / \sin(\beta - \theta)$.

The relationship between the flow deflection angle θ and the shock wave angle β is obtained after trigonometric manipulations of Equations (4.8) and (4.9), resulting in

$$\tan \theta = 2 \cot \beta \frac{(M_1 \sin \beta)^2 - 1}{M_1^2(\gamma + \cos 2\beta) + 2}. \tag{4.10}$$

Equation (4.10) is a classical result known as the θ–β–M relation depicted graphically in Figure 4.5, from which the following observations are made.

- The vertical axis ($\theta = 0$) defines the wave angle resulting from an infinitesimal deflection angle as a function of the Mach number of the upstream flow, which is found from Equation (4.10), so that β equals the Mach angle μ of the oncoming flow.
- For a given value of M_1, the flow deflection has a maximum value for which Equation (4.10) has a solution. For varying flow Mach numbers, this maximum is depicted as a curve suggesting that the corresponding wave angle amounts to approximately 65° for $M_1 > 1.6$. The shock wave will be detached from the corner point in Figure 4.5 if the flow deflection angle is in excess of the top of the constant Mach number curve.
- For $\theta < \theta_{\text{max}}$, there exist two solutions for the wave angle. The smaller value is called the weak-shock solution, the larger one is the strong-shock solution. The strength of the shock wave is defined by the pressure ratio defined by Equation (4.9). Hence, the higher-angle shock wave in Figure 4.5 compresses the air more than the lower-angle wave. The weak-shock solution usually prevails in nature [9].

- The curve for $M_2 = 1$ in Figure 4.5 divides the diagram into regions of supersonic and subsonic downstream Mach number. The region below this curve defines weak-shock solutions with a supersonic downstream flow and a relatively small wave angle.

Strong-shock solutions with a subsonic downstream flow and a large wave angle are situated in the region above the curve for $M_2 = 1$. Since this curve is close to the curve dividing the diagram in regions of different shock strength, it can be stated that in most cases a curved shock planar oblique shock is (relatively) weak with a supersonic downstream flow. The top of each curve for given M_1 identifies the maximum flow deflection for which Equation (4.10) has a solution. Experiments have shown that a flow deflection angle in excess of θ_{max} generates a non-planar shock wave, which is detached from the sharp corner point indicated in Figure 4.5.

4.6 Curved and Detached Shock waves

The distinction between a planar and a curved shock wave is somewhat schematic. Oblique shock waves which are planar in the far field of an object may have one or more curved segments in the near-field flow, as illustrated in Figure 4.6.

Figure 4.6(a). At some distance in front of the depicted blunt-nosed airfoil a curved shock is observed that is known as a bow shock wave. Since at point A the wave is perpendicular to the flow, it represents an example of a strong normal shock wave where the flow deflection is zero and, downstream of point A, the streamline ends in a stagnation point at the airfoil nose. The flow behind the curved shock segment is subsonic in a small patch bounded by a sonic line where $M = 1$. The greatest pressure disturbances invoked by the shock wave are confined to the region between point A and point B located near the sonic line. In this subsonic region the streamlines are curved and the Mach number immediately behind the shock increases outwardly to a maximum in point B, where an almost planar oblique shock has been formed.

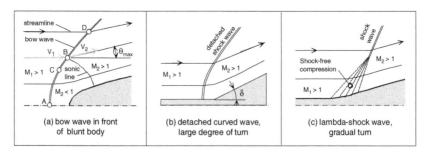

(a) bow wave in front of blunt body

(b) detached curved wave, large degree of turn

(c) lambda-shock wave, gradual turn

Figure 4.6 Detached shock waves.

Figure 4.6(b) depicts a detached shock wave, generated by a wall with a large turn angle $\delta = \theta$ specified on the horizontal axis of Figure 4.5. The flow deflected by the shock wave follows the wall behind the kinked turn. However, the flow deflection angle is in excess of the maximum value for which an attached shock wave is physically possible, with the result that a curved shock wave is formed, which is detached from the sharp corner.

Figure 4.6(c) shows the case where the turn in a wall has a gradual curvature instead of a sharp turn angle. A region of isentropic compression begins were the wall curvature starts with a planar Mach wave at an angle $\sin^{-1}(1/M_1)$, followed by Mach lines with increasing wave angle. The geometry of the fan looks like an upside-down expansion fan. It finishes where the wall becomes planar whereas the straight Mach lines converge into a planar oblique shock wave. The complete flow phenomenon is known as an oblique lambda shock.

4.7 Expansion Flows

Expansion flows are observed in a supersonic airflow expanding round a convex surface that turns away from the oncoming flow, as illustrated in Figure 4.7. The flow follows the surface and it is observed that where it is deflected the pressure, density, and temperature are decreasing while the velocity increases. This type of flow expansion over a well-defined area takes place in an isentropic process and is not sudden as in the case of a shock wave. It is known as a Prandtl–Meyer expansion after the scientists who performed the first research on it in 1908.

Figure 4.7(a) shows an example of an expansion located next to a sharp convex corner, which produces a fan of diverging straight Mach lines in which the curved streamlines diverge. This process as a whole takes place in a continuous expansion region that can be characterized as a continuous succession of Mach

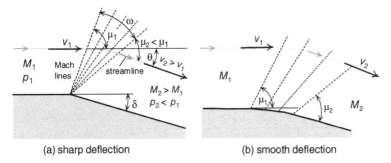

(a) sharp deflection   (b) smooth deflection

Figure 4.7 Expansion of a supersonic flow around a sharp corner or a curved surface deflected away from the oncoming flow.

lines each making the local Mach angle μ with the local flow direction. In the expansion fan, the Mach angle increases from μ_1 for the oncoming flow to μ_2 for the deflected downstream flow. The analysis of the Prandtl–Meyer expansion aims at computation of the flow downstream of the expansion fan for given upstream flow properties and sharp wall deflection angle δ. The flow deflection angle θ increases gradually from zero at the forward Mach line to $\theta = \delta$ at the rearward Mach line. The variation of the infinitesimal deflection of a streamline in the expansion fan is obtained from $d\theta = \sqrt{M^2 - 1}\, dV/V$, where dV relates the change in velocity to the infinitesimal deflection $d\theta$ across a Mach line in the fan. The integration of the deflection angle between M_1 and M_2 leads to the Prandtl–Meyer function

$$\nu(M) = \int_{M_1}^{M_2} \frac{B}{1 + [(\gamma - 1)/2]M^2} \frac{dM}{M} = \sqrt{\frac{\gamma + 1}{\gamma - 1}} \tan^{-1} B \sqrt{\frac{\gamma - 1}{\gamma + 1}} - \tan^{-1} B,$$

$$(4.11)$$

where $B = \sqrt{M^2 - 1}$. The Prandtl–Meyer function is depicted in Figure 4.8. Figure 4.7(b) illustrates that the flow is deflected gradually along a body with a smooth curvature. This process can be treated as the sum of small incremental pressure reductions of a flow initially having a Mach number M_1 and pressure p_1. Induced by successive pressure reductions, the complete expansion process through the Mach wavelets increases the Mach number and decreases the local Mach angle. Since, in contrast to subsonic flows, a supersonic expansion flow is deflected with decreasing pressure, flow separation is less likely to occur. However, it is clear from Equation (4.11) that the deflection angle θ increases over the whole range of Mach numbers. Consequently, for $M \to \infty$, $d\nu(M)/dM \downarrow = 0$

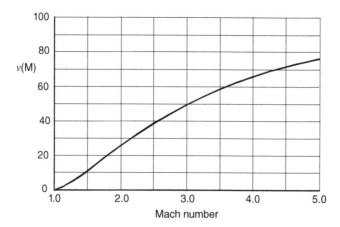

Figure 4.8 The Prandtl–Meyer function defining the supersonic two-dimensional flow properties for an expansion.

and $v(M) \uparrow (\pi/2)(\sqrt{6} - 1) = 4.9\pi/2$. Thus the flow cannot expand from Mach 1 through a turning angle greater than 130° whereas the maximum turning angle amounts to 79° for Mach 5. This is the highest speed for which the Prandtl–Meyer expansion theory is considered to be applicable.

4.8 Shock-expansion Technique

Equation (4.11) or Figure 4.8 can be used to obtain the Prandtl–Meyer function for a given Mach number. If M_1 and M_2 are both given, the deflection angle is obtained from $\theta = \delta = v(M_2) - v(M_1)$. In the more usual case, M_1 and $\theta = \delta$ are given and $v(M_1)$ is solved from Equation (4.11) or Figure 4.8, whereas M_2 is obtained from $v(M_2) = v(M_1) + \theta$. Hence, for a given M_1, M_2 is determined solely by the deflection angle $\theta = \delta$ and if δ increases, so does M_2.

As a secondary result, the pressure downstream of the expansion fan for given upstream flow pressure with the relationship is derived in [9]:

$$\frac{p_2}{p_1} = \left[\frac{2 + (\gamma - 1)M_1^2}{2 + (\gamma - 1)M_2^2} \right]^{\frac{\gamma}{\gamma-1}}. \tag{4.12}$$

To illustrate consequences of the phenomena treated in this chapter, Figure 4.9(a) depicts a schematic flow and pressure distribution on a flat plate at small incidence α to a two-dimensional supersonic flow. The flat plate can be seen as the limiting case of a very thin sharp-nosed symmetrical airfoil section such as the diamond (or double-wedge) shape depicted in Figure 4.9(b). The undisturbed flow in front of the plate is the zone of silence, introduced in Section 4.2.1. Behind the nose, the flow along the upper side is deflected in an expansion fan; the downside flow is compressed by the oblique shock wave emanating from the nose. The situation at the tail mirrors that at the nose, with an oblique shock wave above and an expansion fan below the plate.

The downstream flow behind the tail is deflected upwards and has roughly the same direction as the upstream flow. The streamlines show that there is no downwash behind the plate and the Kutta condition, which is essential for the generation of lift in subsonic flow, does not apply to supersonic flow. Above and below the plate the supersonic flow is parallel to the plate and in inviscid flow the resultant force is normal to the plate. The pressure difference between the upper and lower surfaces exerts a normal force n per unit of span, with lift $l = n \cos \alpha = n$ as the vertical component and drag $d = n \sin \alpha = l \tan \alpha \approx l/\alpha$ as the horizontal component. The magnitude of the (constant) pressure forces on the lower and the upper side of the plate can be calculated with Equations (4.9) and (4.12), respectively. Thus a flat plate at small angle of attack to the flow experiences a lift-dependent

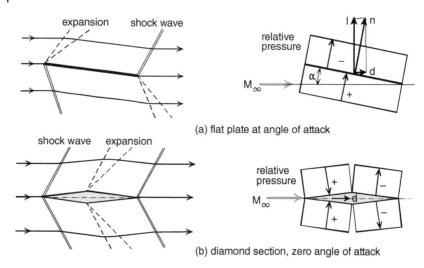

(a) flat plate at angle of attack

(b) diamond section, zero angle of attack

Figure 4.9 Pressure distribution, lift, and drag of sharp airfoils in two-dimensional flow.

drag associated with shock and expansion waves known as wave drag due to lift. This result demonstrates another essential difference to airfoils in subsonic flow, which do not (in theory) experience pressure drag since the suction force acting on the nose and the pressure drag acting on the rear part of the airfoil compensate each other. Application of the shock-expansion technique to airfoil wing sections in two-dimensional flow can be expected to yield accurate results for the pressure distribution and drag, on the provision that skin friction drag is added to the computed wave drag and the sections are thin with sharp leading edges and little camber. The pressure distribution on supersonic airfoils made up of straight-line segments, such as the diamond shape shown in Figure 4.9(b), can be calculated from a combination of the equations for oblique and expansion waves.

This technique has a disadvantage in that it is basically a numerical method that does not yield a closed-form solution for evaluating airfoil performance parameters, such as the lift and drag coefficient. The linearized theory treated in Chapter 5 provides a straightforward method to obtain the pressure distribution on a class of thin airfoils with a more general geometry.

4.9 Leading-edge Delta Vortices

The linearly increasing lift followed by a clearly defined stalled condition as observed for straight wings does not apply to low aspect ratio wings. The airflow over a slender delta wing starts to separate from the leading edge at an

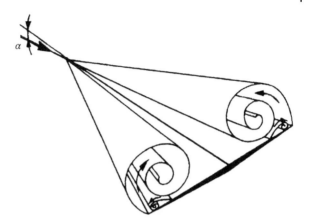

Figure 4.10 Model of vortices above a slender delta wing in subsonic flow.

incidence of just a few degrees. Figure 4.10 illustrates that a powerful and increasingly broadening delta vortex develops above both wing halves, which makes a smaller angle with the approaching flow than the leading edge itself. These vortices create suction forces causing additional lift compared to the linear lift due to the attached flow. Behind the leading edge an area with separated flow and a weak secondary vortex can be distinguished under the delta vortex. Leading edge vortices occur in subsonic as well as supersonic flow. For delta wings in subsonic flow they have the advantage that the lift increases progressively with increasing angle of attack; a favorable phenomenon during the take-off phase of a slender delta wing aircraft. However, strong vortices at the leading edge of a flat delta wing are not favorable a priori for a supersonic vehicle in high-speed flight since they may cause flow separation at a very small angle of incidence of the vehicle, leading to a considerable drag increment, which can only be prevented by giving the wing an appropriately cambered shape. This essential aspect of aerodynamic design forms the subject of Chapter 9.

4.10 Sonic Boom

Supersonic flight has the major drawback that it causes a sonic boom; that is, the result of an observer below the plane sensing the passage of pressure waves caused by a body traveling through the atmosphere at supersonic speed. A supersonic flying airplane is surrounded in the near field by a complex pattern of shock waves and expansion areas. In the far field they are concentrated into a pair of conical pressure waves with an expansion in between. The associated wave pattern stretches over a long distance, most often reaching the ground. The waves are described as a pressure–time history in the form of a sharp pressure rise, followed

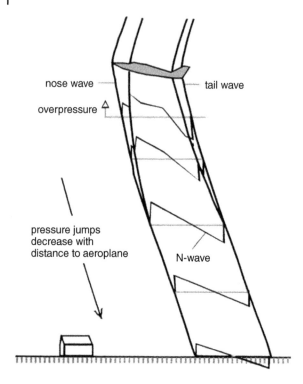

nose wave — — tail wave

overpressure

pressure jumps
decrease with
distance to aeroplane

N-wave

Figure 4.11 Supersonic aircraft surrounded in the near field by a complex pattern of shock waves and expansion areas.

by a steady pressure drop and another sharp pressure rise, known as an N-wave (see Figure 4.11). The waves are reflected by the ground where the pressure fluctuation is about twice the value of the isolated waves. They are perceived within one-tenth of a second or less as a sharp crack or thunder, which is experienced as a serious nuisance.

Bibliography

1 Ashley, H., and Landahl M. *Aerodynamics of Wings and Bodies*. New York: Dover Publications, Inc.; 1965.

2 Clancy, L.J., *Aerodynamics*. New York: John Wiley and Sons; 1975.

3 Kuethe, A.M., and Chow C. *Foundations of Aerodynamics*. 3rd ed. New York: John Wiley and Sons; 1976.

4 Küchemann, D. *The Aerodynamic Design of Aircraft*. 1st ed. Oxford: Pergamon Press; 1978.

5 Bertin, J.J., and Smith M.L. *Aerodynamics for Engineers*. 2nd ed. Englewood Cliffs, NJ: Prentice Hall; 1989.

6 Jones, R.T. *Wing Theory*. Princeton, NJ: Princeton University Press; 1990.

7 Pei Li, and Seebass Richard The Sonic Boom of an Oblique Flying Wing SST, Reprint of paper CEAS/AIAA-95-107; 1995. Available at: http://citeseerx.ist.psu .edu/viewdoc/download?doi=10.1.1.455.7839&rep=rep1&type=pdf

8 Torenbeek, E., and Wittenberg H. *Flight Physics – Essentials of Aeronautical Disciplines and Technology, with Historical Notes.* Springer; 2009.

9 Anderson, Jr., J.D., *Fundamentals of Aerodynamics.* 5th ed. New York: McGraw-Hill; 2010.

10 Panaras, A.G. *Aerodynamic Principles of Flight Vehicles.* Reston, VA: American Institute of Aeronautics and Astronautics; 2012.

11 Ames Research Staff. Equations, Tables and Charts for Compressible Flow. NACA Report 1135; 1953. Available at: https://www.nasa.gov/sites/default/ files/734673main_Equations-Tables-Charts-CompressibleFlow-Report-1135.pdf

12 Jones, R.T. The Minimum Drag of Thin Wings in Frictionless Flow. *J. Aeronautical Sci.* 18 (2):75–81; 1951.

13 Jones, R.T. Theoretical Determination of the Minimum Drag of Airfoils at Supersonic Speeds. *J. Aeronautical Sci.* 19(12); 1952.

14 Jones, R.T. Aircraft Design for Flight below the Sonic Boom Limit. *Canadian Aeronautics Space J.* 20 (5); 1974.

5

Thin Wings in Two-dimensional Flow

Airplane configurations optimized for achieving high aerodynamic efficiency in supersonic cruising flight typically feature a thin and slender wing with sharp trailing edges and a slender fuselage body with a smooth variation of the cross section area between the pointed nose and tail. In order to avoid unfavorable aerodynamic interactions, the arrangement of aircraft components is carefully optimized, and at small angles of attack such a configuration produces weak shock waves, thin boundary layers, and attached flows. An accurate method for computing the pressure distribution on a two-dimensional airfoil is the shock-expansion technique treated in Section 4.8. An alternative and widely used method is the linear theory for thin airfoils, which is applicable to two-dimensional potential flow, and was first published in 1925 by the Swiss scientist J. Ackeret (1898–1981) [11]. This flow model replaces shock waves with Mach waves, disregarding variations in the local Mach number.

5.1 Small Perturbation Flow

Linear theory yields a good approximation of the pressure distribution at locations where flow separation is not dominant and can be used to compute the pressure distribution of thin airfoil sections at a small angle of attack in flows at transonic and low-supersonic Mach numbers. For subsonic as well as supersonic flight at small angles of attack, the two-dimensional flow in which a flying vehicle is immersed can be treated as predominantly isentropic with small perturbations imposed by flow deflections. The analysis of this type of flow is based on the conservation laws of mass, momentum, and energy. Combination of the associated equations yields a set of non-linear partial differential equations that must usually be solved numerically. However, in the absence of viscosity, rotational flows, and shock waves, the external air stream can be represented as a potential flow field. If only small perturbations are manifest in the flow, the velocity potential can

Essentials of Supersonic Commercial Aircraft Conceptual Design, First Edition. Egbert Torenbeek.
© 2020 Egbert Torenbeek. Published 2020 by John Wiley & Sons Ltd.

be modified and solved in terms of a linearized Laplace equation. The properties of such a flow can be computed fairly accurately by means of small perturbation theory. Its solution can then be represented as an approximation for the pressure distribution on the vehicle's surface exposed to the flow. Lift and wave drag are obtained by computing the resultant of the normal forces due to pressure, whereas shear forces are often approximated by means of quasi-empirical methods for predicting friction drag due to viscosity.

The fundamentals of supersonic linearized aerodynamic theory were derived as early as the 1920s. Measurements have proven that in many applications linearized solutions for slender bodies at small incidences to the flow give accurate predictions of experimental results. Linear theory has been successfully applied to computation of the pressure distribution on airfoils during the development of supersonic airplanes in the period 1950–1960, but cannot be used for transonic flow. For certain applications, more complicated solutions obtained from second-order theories or non-linear computational fluid dynamics (CFD) analysis developed since the 1970s have to be preferred. This applies in particular to lift and drag of lifting surfaces intended to realize a significant percentage of the theoretical leading-edge suction (cf. Chapter 9).

5.1.1 Linearized Velocity Potential Equation

Introductions to the velocity potential equation in isentropic flow in which there is no mechanism to start vorticity of the fluid elements can be found in [8] and other publications mentioned in the bibliography of this chapter. A planar wall with a (very small) perturbation immersed in a potential flow field is depicted in Figure 5.1(a). The X-axis is in the direction of the uniform oncoming flow, the Y-axis is normal to it. A velocity potential function Φ can be defined that satisfies the equation $\nabla\vec{\Phi} = \vec{V}$. At an arbitrary point in the flow the local velocity V has

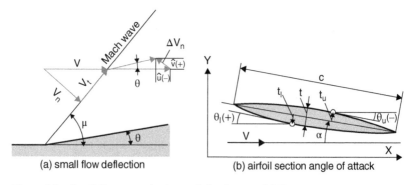

(a) small flow deflection   (b) airfoil section angle of attack

Figure 5.1   Small flow perturbation and circular-arc airfoil geometry.

components $u = V + \hat{u}$ in the X-direction and $v = \hat{v}$ in the Y-direction, where $\hat{u}$ and $\hat{v}$ are called perturbation velocities. Introduction of these concepts into the velocity potential leads to the perturbation velocity potential function $\hat{\Phi}$

$$\Phi = Vx + \hat{\Phi} \quad \text{where} \quad \partial\hat{\Phi}/\partial x = \hat{u} \quad \text{and} \quad \partial\hat{\Phi}/\partial y = \hat{v}. \tag{5.1}$$

In the case of two-dimensional supersonic flow this leads to Laplace's equation,

$$-\beta^2\hat{\Phi}_{xx} + \hat{\Phi}_{yy} = 0 \quad \text{where} \quad \beta \stackrel{\text{def}}{=} \sqrt{M^2 - 1}, \tag{5.2}$$

a non-linear partial differential equation, which mostly cannot be solved analytically. Fortunately, practical solutions have been derived by accepting an approximate solution for the case of slender components of supersonic aircraft in cruising flight. For this case it is assumed that the velocity perturbations $\hat{u}$ and $\hat{v}$ are small in comparison with the oncoming flow velocity and it can be shown that in such cases several terms of the exact velocity potential equation can be ignored relative to the more essential ones. The result is a linear partial differential equation for the perturbation velocities,

$$\beta^2 \frac{\partial \hat{u}}{\partial x} - \frac{\partial \hat{v}}{\partial y} = 0 \quad \text{or} \quad \beta^2\hat{\Phi}_{xx} - \hat{\Phi}_{yy} = 0, \tag{5.3}$$

which appears to be reasonably accurate for slender bodies at small angles of incidence to the flow moving at supersonic speeds not close to the sonic velocity. Equation (5.3) is a hyperbolic differential equation that has a general solution in the form of the functional relation $\hat{\Phi} = f(x - \beta y)$, indicating the property that the velocity potential is constant along straight lines of constant $x - \beta y$. Since a Mach wave has a slope relative to the free flow equal to $\mu = \tan^{-1}(1/\beta)$, the velocity potential is constant along Mach lines. This implies that over a surface featuring a very small ramp angle θ relative to the oncoming flow a Mach wave is generated which is propagated downstream and away from the wall with a slope $dy/dx = 1/\beta$.

5.1.2 Pressure Coefficient

The pressure change of the flow caused by a kink in the surface due to a small perturbation is derived using Figure 5.1(a). Compared with Figure 4.5, the oblique shock with wave angle β is replaced by a Mach wave at an angle μ with the upstream flow. The Mach wave causes a pressure increment when the surface is inclined into the flow, in which case the deflection angle θ is defined as positive. The disturbed velocity behind the Mach wave is the combination of the oncoming flow velocity V and its increments $\hat{u}$ in the X-direction and $\hat{v}$ in the Y-direction. The velocity component normal to the Mach wave $V_n = V \sin\mu$ has changed by ΔV_n, which is decomposed into $\hat{u}$ and $\hat{v}$. For an infinitely small θ it can be assumed that $\Delta V_n \cos \mu = -\hat{u} \cos \mu = \hat{v} = V\theta$. Using $V_n = V \sin \mu$, the pressure change due to the flow deflection follows from Euler's equation

$$\Delta p = -\rho_\infty V_n \Delta V_n = \rho_\infty V^2 \theta \tan \mu,$$ (5.4)

and the pressure coefficient becomes

$$c_p \stackrel{\text{def}}{=} \Delta p/q = \frac{\Delta p}{\frac{1}{2}\rho_\infty V^2} = 2\,\theta \tan \mu = 2\theta/\beta.$$ (5.5)

The sign of θ is positive where the surface is inclined into the free stream flow, leading to an increased wall pressure, and negative where the surface is inclined away from the flow, leading to a reduced wall pressure. Since Equation (5.3) is linear, the pressure distributions due to the airfoil's incidence and the variation of the section thickness and camber along the chord can be calculated separately and then added and the lift, drag, and pitching moment of the airfoil section are found by integration of the resulting pressure distribution along the airfoil contour.

5.1.3 Lift Gradient

The important Equation (5.5) has the consequence that at an arbitrary point of the airfoil surface the pressure coefficient is proportional to the local inclination angle θ which is determined by the angle of attack and the distribution of thickness and camber. Substitution of the airfoil geometry depicted in Figure 5.1(b) into the pressure coefficient according to Equation (5.5) yields the pressure distribution along the airfoil surfaces as follows:

Upper surface: $(c_p)_u = (2/\beta)\, dy_u/dx$,
Lower surface: $(c_p)_l = (2/\beta)\, dy_l/dx$,

where y_u and y_l are the coordinates of the upper and lower surface, respectively. This result implies that the pressure distribution depends only on the airfoil geometry and the angle of attack. As an example, Figure 5.1(b) depicts a circular-arc airfoil at an angle of attack α. A positive α determines a negative pressure coefficient $c_p = -2\alpha/\beta$ on the upper surface and a positive pressure coefficient $c_p = 2\alpha/\beta$ on the lower surface, resulting in a normal force coefficient $c_n = 4\alpha/\beta$.

When carrying out the integration of the pressure distribution along the contour it is observed that the thickness distributions of the upper surface and the lower surface as well as profile camber do not contribute to the lift. Ackeret's theory discovered in [11] suggests that thickness and camber do not improve the lift/drag ratio. The lift coefficient depends on the angle of attack as follows:

$$c_l = c_n \cos \alpha \approx c_n = 4\alpha/\beta \quad \rightarrow \quad dc_l/d\alpha = 4/\beta.$$ (5.6)

In other words, the airfoil experiences lift only due to its incidence to the flow, which is equal to that of a flat plate, and Figure 5.2 compares the lift gradient of an

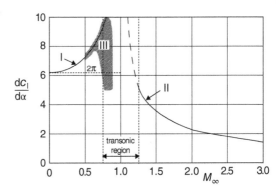

Figure 5.2 The lift gradient of two-dimensional airfoils in subsonic and supersonic potential flow.

airfoil at supersonic and subsonic Mach numbers. The wing of an aircraft in subsonic flight generates the majority of the lift by low pressures acting on the upper wing surface, whereas the zero-lift pressure drag is near-zero on the provision that most of the leading edge suction acting on the nose is fully realized. However, a wing in supersonic flow generates more than half of the lift by a compressive force acting on the lower surface at the cost of an equivalent amount of pressure drag. Figure 5.2 suggests that, according to the Prandtl–Glauert equation, the subsonic lift gradient is considerably higher than the supersonic lift gradient according to Ackeret's theory. On the other hand, the design condition in cruising flight of a supersonic airliner is typically $c_l \approx 0.15$, compared to $c_l \approx 0.50$ for a high-subsonic airplane. In other words, for a specified airfoil lift, both airplane categories require a similar angle of attack.

5.1.4 Pressure Drag

When linearized theory is applied to sharp-edged thin airfoils with a smooth distribution of the upper and lower geometry, the results are qualitatively correct and accurate enough to be used during the initial design stages of a supersonic cruising aircraft. In particular, the predicted pressure distribution along the chord can be useful for the initial stages of aerodynamic design in order to compare different airfoil sections with respect to their contribution to the aerodynamic efficiency of the airfoil. Application of Equation (5.6) yields the coefficient of pressure drag due to lift,

$$(c_d)_l = c_n \, \sin \alpha \approx c_l \, \alpha = (4/\beta)\alpha^2. \tag{5.7}$$

Addition of the pressure drag at zero lift of the upper and lower airfoil parts with thickness t_u and t_l as denoted in Figure 5.1(b) yields the coefficient of pressure drag due to thickness:

$$(c_d)_t = K(t)(4/\beta) \, \tau^2, \tag{5.8}$$

where $\tau \stackrel{\text{def}}{=} t/c$ and the factor $K(t)$ depends on the details of the airfoil contour, as noted in (Equation 5.9). For diamond and lenticular airfoils depicted on Figures 4.9 and 5.1(b) we can write

$$K(t) = 2[(t_u/t)^2 + (t_l/t)^2], \tag{5.9}$$

which amounts to $K(t) = 1.0$ for a symmetric diamond-shape airfoil and $K(t) = 4/3$ for a circular-arc airfoil [9]. The total pressure drag found from addition of the drag due to lift and the thickness drag and the maximum aerodynamic efficiency is obtained for

$$\alpha = \sqrt{K(t)}\,\tau \quad \to \quad (c_l/c_d)_{\max} = 2\sqrt{K_t}\,\tau. \tag{5.10}$$

Combined with the lift according to Equation (5.6) this yields the maximum aerodynamic efficiency

$$c_l/c_d = \left(\alpha + \frac{K(t)\tau^2}{\alpha}\right)^{-1/2}. \tag{5.11}$$

Equation (5.10) demonstrates that the maximum aerodynamic efficiency of a diamond airfoil with $\tau = 0.04$ surrounded by potential flow amounts to $c_l/c_d = 12.5$. This suggests that a two-dimensional sharp-edge thin airfoil experiences a much higher pressure drag for given lift than the pressure drag of a typical low-speed airfoil in subsonic flow.

5.1.5 Symmetric Airfoils with Minimum Pressure Drag

In view of the result from the previous paragraph it is not surprising that much attention has been paid to optimizing the shape of two-dimensional supersonic airfoils. Soon after the development of the linearized supersonic flow theory, it was recognized that a two-dimensional airfoil having minimum pressure drag for a given chord and a given thickness consists of straight lines. It is, however, obvious that in addition to the thickness ratio auxiliary conditions such as the wing volume or some structural requirement may be more important. According to [1], the minimum pressure drag is obtained for the minimum value of the integral

$$I \stackrel{\text{def}}{=} \int_0^c c_p(dy/dx)^2 dx, \tag{5.12}$$

with notations defined in Figure 5.3(a). The auxiliary optimization condition requires the incorporation of an isoperimetric constraint defined by $\int_0^c y^n dx =$ constant, in which the exponent n determines one of the following specifications of the optimization problem and its solution:

Case A for $n = 1$ applies to a constraint on the enclosed sectional area which prescribes the volume per unit span. The result is an airfoil section for minimum

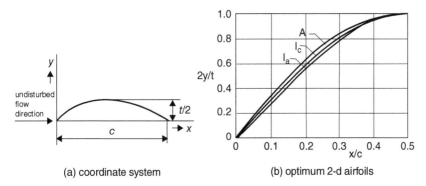

Figure 5.3 Airfoil sections for minimum pressure drag in two-dimensional supersonic flow.

pressure drag defined by the constraint $y = (3Ax/c^3)(c - x)$. This specifies a parabolic arc or (within the limits of linear theory) a circular arc with the associated thickness factor $K(t) = 16/3$.

Case I_c for $n = 2$ prescribes a constraint on the moment of inertia I_c of the contour with respect to the x-axis and the corresponding section has a sinusoidal shape defined by $K(t) = \pi^2/2$ and $y = \sqrt{I_c/c}\,\sin(\pi x/c)$, with the maximum thickness at mid-chord $t = 2\sqrt{I_c/c}$.

Case I_a for $n = 3$ is the same as prescribing the bending stiffness or the torsional stiffness of the structure, determined by the moment of inertia I_a of the sectional area with respect to the x-axis. The analytical solution for the optimum profile is obtained from laborious manipulations and has different solutions for the front and the rear part of the profile. However, the solutions for the optimum thickness and the drag coefficient are simple: $t = 10I_a/c$ and $K(t) = 4.72$.

The resulting airfoil contours are compared in Figure 5.3, suggesting that the three optimum shapes and the corresponding minimum pressure drag are close together. For instance, for specified thickness ratio τ, the pressure drag for case A is 8% higher than for case I_c whereas case I_a has 4.5% less drag than case I_c. And compared with the double-wedge airfoil with the same chord and thickness, the three constrained optimum profiles have between 19% and 1.3% more drag. This does not imply that the double-wedge airfoil is always preferable, since its surface area is a dramatic 30% less than that of the other airfoils.

5.1.6 Total Drag

Ackeret's linear theory applies to inviscid and potential flow and hence skin friction drag should be added to the pressure drag due to lift and thickness. Assuming that the friction drag of a thin airfoil section equals that of a flat plate with the same chord and the same exposed area of the upper and lower surfaces, the total two-dimensional profile drag coefficient amounts to

$$c_{\mathrm{d}} = (4/\beta)[\alpha^2 + K(t)\tau^2] + 2c_{\mathrm{f}}. \tag{5.13}$$

It is readily shown that the maximum aerodynamic efficiency amounts to

$$(c_{\mathrm{l}}/c_{\mathrm{d}})_{\max} = 2(\sqrt{K(t)}\,\tau + \sqrt{2c_{\mathrm{f}}})^{-1/2}. \tag{5.14}$$

Assuming that variation of the flight speed has no effect on the friction drag coefficient, the skin friction drag appears to have a significant effect on the total airfoil drag and may reduce the maximum aerodynamic efficiency by approximately 50%. However, kinetic heating of the boundary layer in high-speed flight causes the skin friction to decrease and the aerodynamic efficiency increases gradually with increasing Mach number. It is also worth noting that according to Equation (5.14) the attainable aerodynamic efficiency of thin, sharp-edged airfoils with thickness ratios between 0.03 and 0.05 in two-dimensional supersonic flow at Mach 2 has the same order of magnitude as Concorde's $(L/D)_{\max}$ as well as published values of supersonic transport airplane projects in Figure 2.4.

5.1.7 Center of Pressure

The pressure coefficient distribution along the chord according to Equation (5.9) is also used to compute the leading-edge pitching moment of a two-dimensional airfoil. The following solution is provided by [9]:

$$c_{\mathrm{m}} = -\frac{2\alpha}{\beta} + \frac{4}{3\beta}\left(\frac{t_{\mathrm{u}} - t_{\mathrm{l}}}{c}\right). \tag{5.15}$$

The center of pressure of a symmetrical airfoil is located at $x_{\mathrm{cp}}/c = c_{\mathrm{m}}/c_{\mathrm{l}} = 0.50$. Since this value is independent of the angle of attack, the aerodynamic center of an airfoil in supersonic flow coincides with the center of pressure: $x_{\mathrm{ac}}/c = 0.50$. This important characteristic differ significantly from the aerodynamic center location of airfoils in subsonic flow, for which the aerodynamic center is located at the quarter-chord location.

5.1.8 Concluding Remarks

It is emphasized that the airfoil properties are derived for two-dimensional wings whereas conditions for three-dimensional wings are quite different. Moreover, linearized potential flow theory is strictly applicable to isentropic supersonic flow with small disturbances and is not valid for transonic Mach numbers. The presence of phenomena such as shock waves, flow separation caused by shock wave/boundary layer interaction, and pockets of subsonic flow are not modeled by linear theory. In spite of these restrictions, results obtained by Ackeret's theory give good insight into the overall effect of varying basic wing shape parameters on the aerodynamic efficiency of airfoils in linearized potential flow.

Different from airfoils in subsonic flow, the thickness drag in supersonic flow is very sensitive to its thickness ratio. Thickness as well as camber contribute to drag and not to lift, and hence a flat plate can be seen as an aerodynamically ideal supersonic airfoil. Obviously, thickness is required to provide a wing with adequate volume and strength and the flat plate does not represent a practical solution. Nevertheless, airfoil sections for supersonic application are much thinner than those for subsonic aircraft: typical thickness ratios are between 3% and 5%.

Bibliography

1 Miele A. (ed). *Theory of Optimum Aerodynamic Shapes*. Academic Press;1965.

2 Ashley, H., and Landahl M. *Aerodynamics of Wings and Bodies*. New York: Dover Publications, Inc.; 1965.

3 Clancy, L.J., *Aerodynamics*. New York: John Wiley and Sons; 1975.

4 Kuethe, A.M., and Chow C. *Foundations of Aerodynamics*. 3rd Ed. New York: John Wiley and Sons; 1976.

5 Houghton, E.L. and Carruthers N.B. *Aerodynamics for Engineering Students*. 3rd ed. Edward Arnold, A Division of Hodder & Stoughton; 1982.

6 Bertin, J.J., and Smith M.L. *Aerodynamics for Engineers*. 2nd ed. Englewood Cliffs, NJ: Prentice Hall; 1989.

7 Jones, R.T., *Wing Theory*. Princeton, NJ: Princeton University Press; 1990.

8 Anderson, Jr., J.D., *Fundamentals of Aerodynamics*. 5th ed. New York: McGraw-Hill; 2010.

9 Panaras, A.G. *Aerodynamic Principles of Flight Vehicles*. Reston, VA: American Institute of Aeronautics and Astronautics, Inc.; 2012.

10 Vos, R., and Farokhi S. *Introduction to Transonic Aerodynamics*. Dordrecht, NL: Springer Science and Business Media; 2015.

11 Ackeret, J. Luftkräfte auf Flügel, die mit groszerer als Schallgeschwindigkeit bewegt werden", Zeitschrift für Flugtechnik und Motorluftschiffahrt. 14 Februar 1925.

12 Jones, R.T. The Minimum Drag of Thin Wings in Frictionless Flow, *J. Aeronautical Sci.* 18 (2); 1951. https://doi.org/10.2514/8.1861

13 Jones, R.T. Theoretical Determination of the Minimum Drag of Airfoils at Supersonic Speeds. *J. Aeronautical Sci.* 19 (12):813–822; 1952.

14 Farren, W.S. The Aerodynamic Art. 44th Wilbur Wright Memorial Lecture. *J. R. Aeronautical Soc.* 60 (547):431–449; 1956

15 Holder, D.W. The Transonic Flow Past Two-Dimensional Airfoils *J. R. Aeronautical Soc.* 68(644):501–516; 1964.

6

Flat Wings in Inviscid Supersonic Flow

Although the wing and the fuselage are the most influential components contributing to aerodynamic drag, in supersonic airplane all main components for optimum efficiency of the complete vehicle must be integrated. For instance: the best airfoil section is not necessarily the section to be used with the optimum planform of the isolated wing, whereas the best planform alone is not necessarily the best to be used for the wing–body combination. In the conceptual design stage of a supersonic cruise vehicle (SCV) this problem may be solved by first conceiving the best wing shape and the best fuselage layout. Significant aerodynamic interaction effects are then identified by means of the area ruling method applied to the complete configuration. However, the design of an optimized SCV wing–body configuration is a complex exercise that will be the subject of Chapters 7 and 8.

The present chapter aims at comparing basic aerodynamic properties of three-dimensional wings based on results generated with linearized theory. The requirement that the disturbances in the flow are infinitesimally small requires that thin and planar wings – also known as flat-plate wings – form the initial subject of the present study. A flat-plate wing is neither cambered nor warped or twisted, a generic shape that has been studied at length because its flow type is illustrative for lifting wings at supersonic speed in general. Pressure drag due to thickness is not taken into account in the present chapter, as opposed to the more comprehensive analysis in Chapter 7 that aims at predicting the pressure drag due to thickness of more realistic wings. Nevertheless, experimental studies have shown that the aerodynamic performance of flat-plate wings with rounded (or "blunt") leading edges can be estimated adequately at small incidences when the flow remains attached over the entire wing surface. Their predicted theoretical performance can be useful for the initial design stage whereas a reliable analysis of aerodynamic performance in more advanced design stages requires the use of an accurate definition of the aircraft geometry and more complex methods to predict the aerodynamic properties by means of non-linear theories and computational fluid dynamics (CFD) methods.

Essentials of Supersonic Commercial Aircraft Conceptual Design, First Edition. Egbert Torenbeek.
© 2020 Egbert Torenbeek. Published 2020 by John Wiley & Sons Ltd.

6.1 Classification of Edge Flows

The lift, drag, and pitching moment of a wing are primarily affected by the Mach number component normal to the leading edge. The following definitions are used in the general case of an arbitrary planform with straight as well as cranked or curved edges:

(a) A supersonic leading edge is a (portion of the) leading edge where the component of the oncoming flow normal to the wing edge is supersonic.
(b) A subsonic leading edge is a (portion of the) leading edge where the component of the oncoming flow normal to the wing edge is subsonic.

Similar definitions apply to supersonic and subsonic trailing edges and side edges. A wing may have supersonic as well as subsonic edges. Wings with only supersonic leading and trailing edges are commonly referred to as having a "simple plan-form" because their aerodynamic analysis is considerably simpler than that of a wing with subsonic edges.

An essential consequence of the present classification is that in the case of a sharp supersonic leading edge there is no interaction between the lower and the upper surface flow. However, if the nose of a subsonic leading edge is blunt, the pressure difference between the lower and upper surfaces may lead to the phenomenon of local subsonic flow and a stagnation point just below the leading edge, causing a high-speed upward airflow and a pronounced suction peak in front of the nose. The overall result of the distributed suction force is known as leading-edge suction, which can bring about a valuable reduction in the pressure drag. This essential subject is treated in Chapter 9. Another consequence for the solution of the three-dimensional potential flow equations is that subsonic trailing edge flow must comply with the Kutta condition, for which the pressure coefficient at the trailing edge at the upper and lower wing surfaces are equal and, hence, the local lift at a subsonic trailing edge is zero.

6.2 Linear Theory for Three-dimensional Inviscid Flow

As noted in Chapter 5, the pressure distribution derived from linearized theory results in a wave drag component, even if shock waves are absent and the flow is assumed to be isentropic. Different from the small-perturbation theory for 2D flow around wing sections, which is based on Equation (5.3), Laplace's equation governing the perturbation velocity potential for three-dimensional flow is a second-order linear partial differential equation of the hyperbolic type,

$$-\beta^2 \hat{\Phi}_{xx} + \hat{\Phi}_{yy} + \hat{\Phi}_{zz} = 0 \quad \text{where} \quad \beta \stackrel{\text{def}}{=} \sqrt{M^2 - 1}. \tag{6.1}$$

Solutions of the linearized three-dimensional theory describe essential aerodynamic properties such as the lift curve slope and the drag due to lift of flat wings with $t/c \ll 1$. However, the linearized theory applies exclusively to isentropic irrotational flow and hence the presence of downwash and a shear layer behind the wing trailing edge must be taken into account by other means.

A classical solution for Equation (6.1) is the conical-flow method that was first proposed by A. Busemann [9]. A conical flow exists in supersonic flow when properties such as velocity components and pressure are invariant along rays emanating from a point where the flow is perturbed. Solutions generated using the conical-flow method were used extensively before the advent of digital computers, although they are still useful to serve as a comparison check to computerized methods. Representative examples of conical flows are observed at the tips of a rectangular wing and at the upper surface of a delta wing with supersonic leading edges.

6.2.1 Flow Reversal Theorems

A remarkable and useful approach to drag minimization problems is to employ certain general theorems that relate the lift and pressure drag distributions in forward and reverse flows. A particular flow reversal theorem states that the pressure drag due to the lift of a wing with given lift distribution is equal to the pressure drag when flying with the same incidence in the opposite direction. Such theorems were first put forward in 1947 by Th. Von Kármán and W.D. Hayes, whereas in 1950 M.M. Munk introduced the concept of a combined flow field concluding that the vortex-induced drag must be the same in forward and reverse flow. Flow reversal theorems have also been used by R.T. Jones to derive criteria for identifying configurations of minimum drag [1].

6.2.2 Constant-chord Straight Wings

Early applications of linear flow theory were developed during the early 1980s for flat plate wings with constant chord. Geometric parameters of the wing to be studied were its aspect ratio and the angle of sweep. In particular the rectangular wing has been studied by many investigators. The flat rectangular wing depicted in Figure 6.1 is surrounded by supersonic flow normal to its leading and trailing edges. The pressure at any point P on the wing is affected only by disturbances generated at points within the Mach cone emanating upstream from P. Since perturbations occur only at the two tip leading edges, the wing pressure distribution is affected in triangular regions within the Mach cones emanating from the tips.

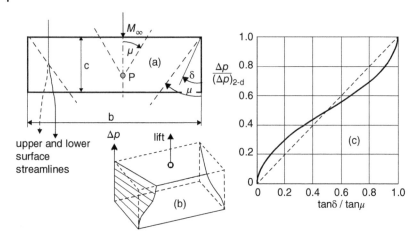

Figure 6.1 Pressure distribution and streamlines for a flat rectangular wing. (a) Conical flow regions and upper and lower surface streamline deflection at the tips. (b) Three-dimensional pressure distribution. (c) Variation of the pressure difference inside the tip Mach cones.

Figure 6.1(c) depicts the variation of lift produced by the tip Mach cones, a classical example of conical flow defined by

$$\frac{\Delta p}{(\Delta p)_{2\text{-}d}} = \frac{2}{\pi}\sin^{-1}\sqrt{\frac{\tan \delta}{\tan \mu}}, \tag{6.2}$$

where δ and μ are defined in Figure 6.1(a). Equation (6.2) applies to isolated rectangular wing tips, and the lift per unit area generated by the region inside each tip Mach cone is 50% of the lift per unit area of a two-dimensional wing. The remainder of the wing between the tip Mach cones is not influenced by any upstream perturbation and hence the pressure distribution in this region can be derived from the two-dimensional theory exposed in Chapter 5.

From Equation (6.2) it can be derived that, relative to a two-dimensional wing with the same incidence to the flow, the lift and the pressure drag due to lift ΔC_{D} of a rectangular wing are reduced by a factor $1/(2\beta A)$, where the aspect ratio is defined as $A \stackrel{\text{def}}{=} b/c$. Alternatively, the lift gradient and the coefficient of pressure drag due to lift can be written as follows:

$$\beta C_{\mathrm{L}_\alpha} = 4[1 - 1/(2\beta A)] \quad \text{and} \quad \Delta C_{\mathrm{D}} = 1/C_{\mathrm{L}_\alpha}. \tag{6.3}$$

Equation (6.3) suggests that the effect of varying the aspect ratio of a rectangular wing in supersonic flow is less essential compared to subsonic flow. It can also be shown that the center of pressure of a rectangular wing shifts forward from the mid-chord point in two-dimensional supersonic flow over a distance

$$\Delta x_{\mathrm{ac}}/c = 6(2\beta A - 1)^{-1}. \tag{6.4}$$

Inspection of this result indicates that the effects of the finite aspect ratio on the aerodynamic properties of a rectangular wing are not negligible for slender wings that have an aspect ratio A typically less than one. However, the following restrictions on the validity of the present theory must be taken into account:

- Equations (6.2) through (6.4) are valid on the provision that the Mach cones emanated by the tips do not intersect at the wing plane, which appears to be case for $\beta A \geq 2$. If this condition is not satisfied the analysis must be extended to the case that the tip cones are interacting. The pressure distribution in the region of overlap is then determined by adding the pressures inside each tip cone and subtracting their sum from the pressure field as determined by the two-dimensional linearized theory.
- Inside the tip Mach cones there is an exchange of pressure between the lower and the upper surfaces leading to an upward flow around the tips. Figure 6.1(a) shows that the streamlines are deflected inwards at the lea-side surface and outwards at the lower surface. Inside the Mach cones the discontinuity of lateral flows behind the trailing edge causes a shear layer behind the outboard wing. The drag associated with this phenomenon may be treated with a non-linear theory including viscosity, but its effect is insignificant unless the wing has a very small aspect ratio so that it must be treated as a slender wing. In that case a strong conical vortex will develop at the side edges. Such a wing is unlikely to be a candidate for application for an efficient SCV.

6.2.3 Constant-chord Swept Wings

Since the lift/drag ratio of a rectangular wing at supersonic speeds appeared to be even worse than that of two-dimensional wing sections, it was soon realized that application of leading-edge sweep is required to obtain acceptable high-speed flight performances. Consequently, swept wings became the subject of many investigations and applications to a generation of transonic military aircraft. The flow around a swept wing with constant chord length exhibits a pressure distribution that is highly dependent on the flight Mach number, as illustrated in Figure 6.2.

(a) If the wing is sufficiently swept so that the leading and trailing edges are subsonic, the flow around the wing has everywhere a subsonic character. Even if the flight speed is supersonic, a wing with subsonic leading edges in the design (cruising) condition may feature acceptable aerodynamic design properties. For instance, the nose of a wing with subsonic leading edge can be smoothly rounded without generating a drag-producing bow wave in front of a blunt nose in two-dimensional flow (Figure 6.1).

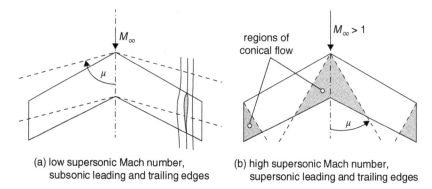

(a) low supersonic Mach number,   (b) high supersonic Mach number,
    subsonic leading and trailing edges   supersonic leading and trailing edges

Figure 6.2 The flow around swept wings with constant chord.

(b) If the flow Mach number is high so that the leading and trailing edges are supersonic, Mach cones not only originate at the tip leading edges but also at the the wing apex. In this situation the center portion is covered by a Mach cone in which the trailing edges exhibit upward oblique shock waves and downward expansion waves. Compared to a rectangular wing with the same span and chord, the three regions with reduced pressure together are larger and their effects on the lift curve slope, wave drag due to lift, and pitching moment are more pronounced.

As the flight speed is increased to higher Mach numbers, the angle of sweep has to be increased in order to keep the leading edge subsonic and it becomes increasingly difficult to keep the flow perturbations small. Experiments have shown that the associated flow separation over the upper wing surface are highly non-linear, with the effect that the theoretical benefits of sweep-back are not always attained in practice. Elimination of separated flow in the design condition can be achieved by blending the effects of leading- and trailing-edge sweep-back angles, thickness distribution, leading-edge nose radius, and camber and twist variation along the span. However, similar to the rectangular wing, the constant-chord swept wing with supersonic leading edges has unfavorable aerodynamic properties for application to a supersonic cruise vehicle.

6.3 Slender Wings

Classical supersonic wing theory indicates that, in order to achieve low drag in cruising flight, the leading edge must have an angle of sweep greater than the Mach angle[1]. A wing is considered as slender if it has a small span compared to its length;

---

1 In fact, it is a sensible aim to keep the whole aircraft well within the Mach cone from its nose.

in other words, slender wings have a low aspect ratio[2]. The first generations of (military) supersonic aircraft often used thin rectangular and low aspect ratio straight or swept-back wings. One aim of their aerodynamic design has been to find shapes that would combine adequate flight performances as well as good flying qualities. This geometry ruled out airfoils with blunt leading edges resulting in poor supersonic performance, whereas wings with sharp leading edges cannot support the classical attached flow required to fly at low speeds, unless variable-geometry devices such as leading-edge flaps are applied. These requirements lead to the concept of a configuration with sufficiently swept and aerodynamically sharp leading edges where separation is fixed under all flight conditions. For application at cruise speeds higher than Mach 2 a delta wing is geometrically slender; that is, the aspect ratio should not be greater than approximately one. Moreover, the leading edges are attachment lines at one flight condition.

6.4 Delta Wing

A classical planform applied to supersonic airplanes is the delta wing, which allows application of a slender shape with acceptable aerodynamic properties at high Mach numbers as well as in subsonic flight. The delta wing concept was developed by German engineers during the second World War and has often been used for military applications, in particular because delta wings have excellent aerodynamic properties at transonic and supersonic speeds. The basic triangular delta wing shape depicted in Figure 6.3 has a low aspect ratio (typically $1.0 < A < 3.0$) with highly swept leading edges, zero trailing-edge sweep and zero taper ratio[3].

The character of the flow past a flat delta wing is characterized to a large extent by the leading edge flow parameter $m = \tan \gamma / \tan \mu$, with γ denoting the complement of the leading edge sweep angle Λ_{le} as depicted in Figure 6.3. For a sonic leading edge the flow parameter $m = 1$, whereas for a supersonic leading edge $m > 1$ and for a subsonic leading edge $m < 1$. The relation between the wing area S, the aspect ratio A, and the leading edge sweep angle Λ_{le} in Figure 6.3 holds for a pure delta wing and is not exactly valid for modified versions of the basic delta shape, treated in Section 6.6. Since the flow around slender wings in general is primarily determined by the leading edge sweep angle, $m = \beta \cot \Lambda_{le}$ is the preferred definition for the flow parameter. Moreover, it is also valid for arrow wings

---

2 The term slender wing can lead to misunderstanding: a slender wing has a low aspect ratio from the aerodynamic point of view, whereas a high aspect ratio wing is considered to have a slender structure.
3 The slender wing with near-triangular planform and sharp leading edges was the most prominent aspect of Concorde's wing.

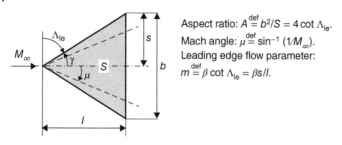

Aspect ratio: $A \overset{\text{def}}{=} b^2/S = 4\cot\Lambda_{\text{le}}$.

Mach angle: $\mu \overset{\text{def}}{=} \sin^{-1}(1/M_\infty)$.

Leading edge flow parameter:

$m \overset{\text{def}}{=} \beta\cot\Lambda_{\text{le}} = \beta s/l$.

Figure 6.3 Basic delta wing geometry and definitions of flow parameters.

with straight leading and trailing edges for which the definition $m = \beta A/4$ does not apply.

6.4.1 Supersonic Leading Edge

Figure 6.4 depicts a delta wing at an incidence α to the oncoming flow with speed of such a magnitude that the component of the speed normal to the leading edge V_{n} exceeds the sonic velocity; that is, $m > 1$. As explained in Section 6.2, a pressure disturbance generated on a supersonic leading edge only affects the region within the Mach cone emanating from the point where the disturbance is generated. Consequently, point P on the leading edge only experiences the influence of upstream pressure disturbances produced within the Mach cone mirrored forward from it. This cone extends within the zone of silence in front of the leading edge and hence no point on the wing produces a disturbance that influences the flow at any point located at the leading edge.

A pure delta wing has straight leading edges and the only disturbance affecting its pressure distribution is located at the vertex A, which emanates a disturbed flow

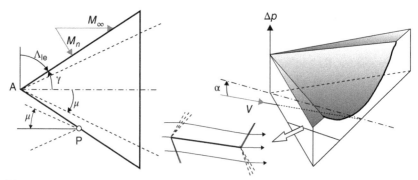

(a) supersonic leading edge delta   (b) schematic flow pattern and pressure distribution

Figure 6.4 Geometry and pressure distribution of a flat delta wing with straight supersonic leading edges.

field inside the conical Mach wave. The flow in front of this flow field, sketched in the inset figure, is similar to a supersonic two-dimensional flow, for which the constant pressure difference between the upper and lower surface follows from Ackeret's theory, when applied to M_n. In this region, the constant lift per unit area proves to be larger than for a two-dimensional wing with the same Mach number and incidence. Within the Mach cone emanating from A there is a conical-flow field as defined in Section 6.2.2.

Figure 6.4(b) indicates that in this region the pressure difference has a minimum in the plane of symmetry, and no points on the trailing edge experience the influence of adjacent points. The Kutta condition is not satisfied at the trailing edge and hence the pressure difference and the lift disappear abruptly, indicating the presence of shock waves and expansion regions as sketched. If the pressure distribution for the region within the Mach cone is computed with the conical-flow method it is concluded that the integrated pressure force on the wing is less than that according to Ackeret's theory.

Instead of computing the flow with the conical-flow method, the flow reversal theorem can be applied, with the implication that the normal pressure force of the delta wing with supersonic leading edges is equal to the pressure force on the same delta wing in reversed flow. Accordingly, the trailing edge then becomes a straight supersonic leading edge in the reverse flow, producing a constant normal pressure on the upper and lower wing surface and hence the flat delta wing with supersonic leading edges generates the same average lift per unit of area as a two-dimensional flat plate at the same incidence to the flow approaching from behind,

$$C_L = \frac{4\alpha}{\sqrt{M_\infty^2 - 1}} = \frac{4\alpha}{\beta}. \tag{6.5}$$

The lift gradient is thus obtained from $\beta C_{L_\alpha} = 4$, whereas the induced drag is obtained by using Figure 6.3 as follows:

$$\Delta C_D = C_N \sin \alpha \approx C_L \alpha = C_{L_\alpha} \alpha^2 = \beta/4 C_L^2 \quad \rightarrow \quad \Delta C_D = \beta C_L^2/4. \tag{6.6}$$

The induced drag can also be written as $\Delta C_D (C_L^2/\pi A)^{-1} = \pi m$, indicating that accepting a supersonic leading edge brings about a considerable induced drag penalty compared to the ideal minimum induced drag at subsonic speeds.

6.4.2 Subsonic Leading Edge

When a delta wing is placed in a lower-supersonic airflow, the Mach angle increases and the Mach waves emanating at the wing vertex rotate towards the leading edge. For the situation in Figure 6.5, the speed is low enough to make the Mach angle μ larger than the angle γ and the flow parameter m becomes less than one. The velocity component normal to the leading edge is now subsonic

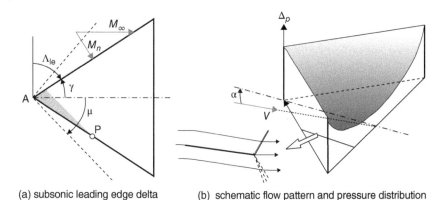

(a) subsonic leading edge delta   (b) schematic flow pattern and pressure distribution

Figure 6.5 Geometry and pressure distribution of a delta wing with subsonic leading edges.

and the entire wing is inside the Mach cone emanating from the vertex. In this situation the wing has a subsonic leading edge and a supersonic trailing edge, whereas point P on the leading edge experiences the influence of pressure disturbances emanated by all points within its (mirrored) upstream Mach cone originating from the shaded area of the wing. Although the flow past the leading edge is supersonic, its character is determined by the subsonic component M_n and, similar to Figure 6.2(a), the leading edge of a delta wing is surrounded by flow from the stagnation point below the nose, generating a forward suction force the nose which effectively acts as a reduction of the drag[4].

The lift gradient of a delta wing with subsonic leading edges $m < 1$ is obtained from slender wing theory [5]:

$$\beta C_{L_\alpha} = \frac{2\pi m}{E'(m)}, \qquad (6.7)$$

where $E'(m)$ denotes the elliptic integral of the second kind with modulus m, defined as follows:

$$E'(m) = \int_0^{\pi/2} [1 - (1 - m^2)\sin^2\phi \ d\phi]^{1/2}. \qquad (6.8)$$

The following accurate approximation for $E'(m)$ is proposed in [19]:

$$E'(m) = 1 + (\pi/2 - 1)m^\eta \quad \text{where} \quad \eta = 1.226 + 0.15\pi(1 - \sqrt{m}) \qquad (6.9)$$

Figure 6.6(a) suggests that for slender delta wings with $m < 0.2$ the lift gradient approaches $C_{L_\alpha} = \pi A/2$ and it is worth noting that the lift gradient of a slender

---

4 Strictly, if a flat wing has a sharp nose, the leading edge flow will separate and the suction force cannot develop, but practical (very thin) wings have some degree of roundness at which a fraction of the suction force may be realized.

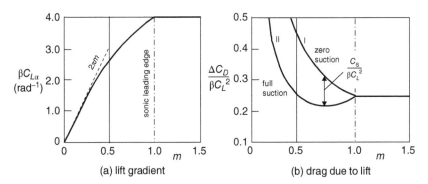

Figure 6.6 Lift gradient and induced drag flat delta wings according to linearized theory.

wing in subsonic flow is identical to its counterpart with the same aspect ratio in supersonic flow.

The induced drag due to lift for a subsonic leading edge with zero leading edge suction is derived from Equation (6.7),

$$\frac{\Delta C_D}{\beta C_L^2} \approx \frac{1}{\beta C_{L_\alpha}} = \frac{E'(m)}{2\pi m}. \tag{6.10}$$

Equation (6.10) is depicted in Figure 6.6(b) as curve I. If the full leading edge suction can be realized, the induced drag coefficient is reduced by the suction coefficient,

$$\frac{C_S}{\beta C_L^2} = \frac{\sqrt{1 - m^2}}{4\pi m}, \tag{6.11}$$

corresponding to the minimum obtainable induced drag

$$\frac{\Delta C_D}{\beta C_L^2} = \frac{E'(m)}{2\pi m} - \frac{\sqrt{1 - m^2}}{4\pi m}, \tag{6.12}$$

depicted in Figure 6.6(b) as curve II. As mentioned before, a flat-plate delta wing is a theoretical concept that will not generate leading-edge suction. Prediction of the obtainable leading-edge thrust for practical wing shapes with finite thickness, rounded noses and/or camber is an essential part of aerodynamic design to be treated in Chapter 9.

In the present context reference is made to comprehensive publications such as [19] and [22], whereas the following provisional observations can be useful in the conceptual design stage.

- The favorable effect of suction increases with decreasing m. However, if no leading-edge suction is realized, the drag due to lift is doubled for $m < 0.5$.

- In the design condition the leading edge flow parameter m is usually in excess of 0.5 and Figure 6.6(b) shows that for $0.5 < m < 1.0$ the drag due to lift is significantly lower for a delta wing with full leading edge suction compared to one with supersonic leading edges. On the provision that the leading edge suction is fully realized, the minimum value of $\Delta C_D / C_L^2$ is obtained when $m = \cot \Lambda_{le} \approx 0.75$, corresponding to Mach 0.75 normal to the leading edge.
- In subsonic airflow, the induced drag is inversely proportional to the wing aspect ratio, the essential reason why all subsonic airliners have a high-aspect-ratio wing. However, if the equations for linear theory are applied to delta wings with different aspect ratios it is observed that, for flight speeds between Mach 1.2 and 2.0, variation of the aspect ratio has little influence on the drag due to lift for delta wings with full leading edge thrust. This explains to some extent why supersonic cruising aircraft have a low aspect ratio[5].

6.5 Arrow Wings

Derived from the delta shape, the arrow(head) wing was conceived during the 1970s in the framework of the SCAR research program. The arrow wing is created by a trailing edge cut-out that deletes part of the delta wing where the generation of lift is less effective, as depicted in Figures 6.4 and 6.5. In the case of notched trailing edges shown in Figure 6.7 the wing geometry is defined by the notch ratio a. For a planform area S the arrow shape enables the wing's

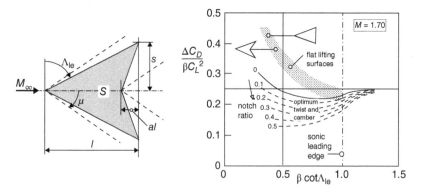

Figure 6.7 Geometry of the flat arrow wing and effect of the notch ratio on the induced drag of arrow wings (after [15]).

---

5 The oblique wing aircraft configuration described in Chapter 10 forms an exception to this rule.

length l to be increased, resulting in a reduced pressure drag due to thickness. However, a more prominent advantage of the arrow wing is in the area of induced pressure drag. This is illustrated in Figure 6.7, although it is emphasized that most of the improvements relative to the delta wing are due to twist and camber. This important aspect is subject to treatment in Chapter 9.

A comparison between delta and arrow wing configurations in [17] designed for cruising at Mach 2.5 shows that the arrow wing design has a lower zero-lift drag than a delta wing design, which is attributed to a higher wave drag for the supersonic leading edge delta wing and a lower slenderness ratio of its equivalent body. Compared to a delta wing configuration with the same leading edge sweep, the drag reduction of a typical arrow wing amounts to 6.5% for the zero-lift drag and 15% for the induced drag, resulting in 12% improvement of the aerodynamic efficiency. The overall mission impact for the configuration discussed in [15] is a range improvement of 1,000 km due to the L/D advantages[6].

6.6 Slender Delta and Arrow Wing Varieties

For subsonic as well as supersonic leading edges, two-dimensional and conical flow regions have the property that the pressure is constant along straight lines through the wing vertex. The numerical analysis in [14] confirms that, independent of the flow parameter m, the center of pressure of a pure delta wing coincides approximately with its center of area located at 2/3 of the root chord downstream from the vertex. For supersonic flow this point is also the aerodynamic center. The backward shift of the aerodynamic center of an airfoil in subsonic flow from 25% of the chord behind the leading edge to the more rearward location at supersonic speed has a significant effect on the stability and control of the aircraft. Accordingly, several modifications of the basic wing have been developed aiming at bringing the aerodynamic center at supersonic speeds more forward.

Figure 6.8 illustrates a variety of modifications intended to improve the aerodynamic properties of the delta wing.

(a) The clipped delta wing has a (small) taper ratio in order to eliminate the inefficient narrow-chord tip region of the pure delta wing.
(b) The cranked leading edge of the double delta wing is mainly used to reduce the aerodynamic center movement in the transition from subsonic to supersonic flight.
(c) The arrow wing with straight leading edges is described in Section 6.5.

---

6 In principle, this result must be adjusted for the weight fraction differences between the delta and the arrow wing.

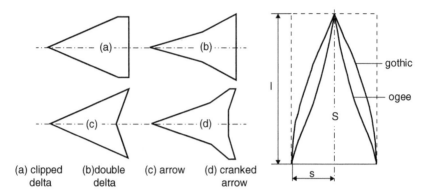

Figure 6.8 Varieties of delta and arrow wings with straight leading edges and delta wing modifications with curved leading edges.

(d) The cranked arrow wing is a further refinement, having an inboard wing with a subsonic leading edge and a (very thin) outboard wing with a supersonic leading edge.

The cranked arrow can profit from the attached flow at the highly swept leading edge. Leading-edge flaps or a variable nose camber may be used for achieving high aerodynamic efficiency in the operational range of Mach numbers and incidences relative to the flow. This hybrid wing concept is considered to be a promising candidate for a future supersonic commercial transport.

The geometry of slender wings is not restricted to delta shapes with straight leading edges. Figure 6.8 depicts two classes of curved leading edges with streamline tips described in [2].

- The gothic wing has a curved leading edge that remains convex along its length. Its shape is defined as $s(x)/s = 2(x/l)(1 - x/l)$, where the local semi-width is defined as $x(s)$.
- The ogival delta has a leading edge with an inflection point and a shape defined as $s(x)/s = 0.8(x/l) + 0.6(x/l)^4 - 0.4(x/l)^8$.

All slender wings have a pointed apex and the flow around the neighborhood of the apex is approximately conical. This maximizes the suction force on the nose and distributes the lift favorably in a lateral direction, thereby minimizing the drag due to lift.

Bibliography

1 Ashley, H., and Landahl M. *Aerodynamics of Wings and Bodies.* New York: Dover Publications, Inc.; 1965.

2 Küchemann, D. *The Aerodynamic Design of Aircraft*. Oxford: Pergamon Press; 1978.

3 Jones, R.T. *Wing Theory*. Princeton, NJ: Princeton University Press; 1990.

4 Anderson, Jr., J.D. *Fundamentals of Aerodynamics*. 5th ed. New York: McGraw-Hill; 2010.

5 Jones, R.T. Properties of Low-Aspect Ratio Wings at Speeds Below and Above the Speed of Sound. NACA Report 835; 1946. Available at: http://naca.central .cranfield.ac.uk/reports/1946/naca-report-835.pdf

6 Jones, R.T. Wing Plan Forms for High-Speed Flight. NACA TN 863; 1947. Available at: https://ntrs.nasa.gov/archive/nasa/casi.ntrs.nasa.gov/19930091936 .pdf

7 Jones, R.T. Estimated Lift-Drag Ratios at Supersonic Speed. NACA TN 1350; 1947. Available at: https://digital.library.unt.edu/ark:/67531/ metadc55497/

8 Bonney, E.A. Aerodynamic Characteristics of Rectangular Wings at Supersonic Speeds. *J. Aeronautical Sci.* 14(2):110–116; 1947.

9 Busemann, A. Infinitesimal Conical Supersonic Flow. NACA TM-1100; 1947. Available at: http://naca.central.cranfield.ac.uk/reports/1947/naca-tm-1100.pdf

10 Jones, R.T. The Minimum Drag of Thin Wings in Frictionless Flow. *J. Aeronautical Sci.* 18(2):75–81; 1951.

11 Jones, R.T. Theoretical Determination of the Minimum Drag of Airfoils at Supersonic Speed. *J. Aeronautical Sci.* 19 (12):813–822; 1952.

12 Küchemann, D. *Some Considerations of Aircraft Shapes and their Aerodynamics for Flight at Supersonic speeds*. Advances in Aeronautical Sciences 3 London: Pergamom Press; 1961. p. 221.

13 Stanbrook, A., and Squire L.C. Possible Types of Flow at Swept Wing Leading Edges. *The Aeronautical Quarterly*, 15(1):72–82; 1964.

14 Middleton, W.D., and Carlson H.W. A Numerical Method for Calculating the Flat-Plate Pressure Distributions on Supersonic Wings of Arbitrary Planform. NASA Technical Note D-2750, January; 1965.

15 Baals, D.D., Robins A.W., and Harris R.V. Aerodynamic Design Integration of Supersonic Aircraft. *J. Aircraft.* 7(5); 1968.

16 Carlson, H.W., and Miller D.S. Numerical Methods for the Design and Analysis of Wings at Supersonic Speeds. NASA Technical Note D-7713, December; 1974.

17 Wright, B.R., Bruckman F. and Radovcich N.A. Arrow Wings for Supersonic Cruise Aircraft. *J. Aircraft*, 15(12):829–836; 1978.

18 Kulfan, R.M., and Sigalla A. Real Flow Limitations in Supersonic Airplane Design AIAA Paper 78-147, January; 1978. https://doi.org/10.2514/6.1978-147

19 Carlson, H.W., and Mack R.J. Estimation of the Leading-Edge Thrust for Supersonic Wings of Arbitrary Planform. NASA Technical Paper 1270, October; 1978.

20 Squire, L.C. Experimental Work on the Aerodynamics of Integrated Slender Wings for Supersonic Flight. *Progr. Aerospace Sci.* 20(1);1–96; 1981.

21 Mattick, A.A., and Stollery J.L. Increasing the Lift: Drag Ratio of a Flat Plate Delta Wing. *Aeronautical J.* 85(848):379–386; 1981.

22 Wood, R.M., and Miller D.S. Impact of Airfoil Profile on the Supersonic Aerodynamics of Delta Wings. *J. Aircraft* 23(9):695–702; 1986.

23 Wood, R.M. Supersonic Aerodynamics of Delta Wings. NASA TP 2771, March; 1988.

7

Aerodynamic Drag in Cruising Flight

Aerodynamic drag in horizontal flight is usually expressed in terms of aerodynamic efficiency $L/D = C_L/C_D$. This term suggests that, similar to energy efficiency, L/D would have a value between zero and one, which could be confusing to engineers who are not familiar with analysis of flight mechanics. For instance, Concorde's aerodynamic efficiency in cruising flight at Mach 2 was $L/D = 7.5$, whereas for typical contemporary passenger transport and long-range jetliners, L/D would have an order of magnitude between 15 and 18, typically. In deriving the sensitivity of L/D to the variation of geometric variables it is often attractive to use the term glide ratio C_D/C_L since C_L is usually the independent variable. This leads to expressions which are simpler to solve for optimum independent variables affecting the drag. Skin friction drag depends predominantly on the area exposed to the surrounding flow, which is usually called the "wetted area", whereas vortex-induced drag is primarily a function of wing span. However, designers of a supersonic cruising vehicle (SCV) have to contend with substantial wave drag associated with the volume of the primary aircraft components. The SCV configuration designer must therefore get used to significantly lower aerodynamic efficiencies in cruising flight than typical values for subsonic jetliners.

Studies developed in the framework of second generation SCV development suggested that (around the year 2000) $L/D \approx 10$ will be achievable for cruising at Mach 2, typically. However, the next generation of subsonic jetliners is expected to achieve up to $L/D \approx 25$ and to be able to compete with them a second generation SCV will have to generate an aerodynamic efficiency that is considerably in excess of $L/D = 10$, as indicated in Figure 7.2.

7.1 Categories of Drag Contributions

Figure 7.1 illustrates how the aerodynamic force acting on an SCV can be decomposed according to several schemes associated with the method used to derive

Essentials of Supersonic Commercial Aircraft Conceptual Design, First Edition. Egbert Torenbeek.
© 2020 Egbert Torenbeek. Published 2020 by John Wiley & Sons Ltd.

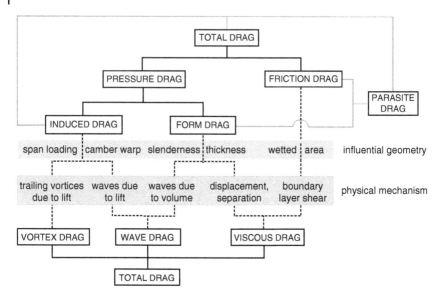

Figure 7.1 Scheme for decomposing the drag of a supersonic flight vehicle.

the separate components. The most fundamental approach is decomposing the drag into distributed normal pressure forces and shear forces on the plane's area exposed to the flow. However, from the flight performance point of view, the basic force components are lift acting normal and drag acting tangential to the direction of flight.

Different from subsonic flight, the zero-lift drag and drag due to lift are increased in supersonic flight by pressure drag associated with the existence of pressure waves.

- The zero-lift drag D_0 summarizes the skin friction drag D_F and the wave drag due to volume D_{WV}. Both are treated as practically independent of the airplane incidence to the flow.
- The induced drag ΔD – also denoted as drag due to lift – combines the wave drag due to lift D_{WL} and the vortex-induced drag D_{VL}.

In terms of coefficients, the airplane total drag is decomposed into a constant term $C_{D_0} = C_{D_F} + C_{D_{WV}}$ and a term $\Delta C_D = C_{D_{WL}} + C_{D_{VL}}$ which is considered to be proportional to C_L^2. Consequently, the four basic drag terms to be treated in the present chapter are

$$C_D = C_{D_0} + \Delta C_D = C_{D_F} + C_{D_{WV}} + C_{D_{WL}} + C_{D_{VL}}. \tag{7.1}$$

This parabolic relationship is sufficiently representative for the initial design stage of an SCV with a planar wing since several minor drag components cannot be analyzed until sufficiently detailed information is available on the shape of the surface exposed to the flow. This simplification does not neglect the fact that in reality the wing shape should be designed so that the minimum drag is achieved in the cruise condition. In particular, non-planar wings are applied to minimize the drag due to lift by selecting properly cambered wing configurations, including application of wing warping and twisting.

7.1.1 Miscellaneous Drag Terms and the Concept Drag Area

The following drag contributions not explicitly incorporated in Figure 7.1:

- The scheme of Figure 7.1 does not explicitly depict a nominal force component associated with the low pressure acting on rounded wing leading edges denoted as leading-edge suction. This phenomenon results in leading edge thrust, which is usually treated as a drag reduction, an essential subject treated in Chapter 9.
- Trim drag is associated with balancing the aircraft in level flight. For a given location of the center of gravity, the trim drag depends on the longitudinal distribution of lift and hence it can be expressed as an increment of the induced drag.
- Drag associated with the operation of the installed power plant, an effect that may be taken into account in the initial design stage as a reduction of the propulsive efficiency[1].

By means of dimensional analysis it can be shown that the drag of a body moving with speed V in a fluid with density ρ is proportional to the body reference area S and the ambient flow dynamic pressure $q = \frac{1}{2}\rho V^2 = \frac{1}{2}\gamma p M^2$. This relationship is often applied by introducing the drag $C_D S \overset{\text{def}}{=} D/q$. Using the drag area instead of the drag coefficient avoids confusion since the choice of the reference area may be arbitrarily chosen and therefore not always adequately defined.

7.1.2 Analysis Methods

Two basic theoretical methods are distinguished to analyze the drag of a vehicle flying at supersonic speed.

(1) The near-field theory involves computation of normal and tangential forces on the aircraft's three-dimensional surfaces exposed to the surrounding flow. This approach, also known as the singularity method (or panel) method, requires the vehicle's surface to be modeled by a large number of quadrilateral elementary panels to simulate their effect on the flow field. This approach involves

---

1 Drag data quoted for existing aircraft do not always comply with the engine integration aspect and one must be careful in comparing aircraft overall drag data with airframe drag data.

an accurate representation of the aircraft's external surface, division of this surface into small elements and computation of the elemental pressure and skin friction forces normal and tangential to the local surface. All elemental forces are decomposed into components normal to and in the direction of the oncoming flow and their summation yields the overall lift and drag forces.

(2) The far-field theory involves momentum considerations across a very large cylindrical volume encompassing the complete airplane and its surrounding flow phenomena. This approach can be used to compute wave drag due to the volume of aircraft components exposed to the flow, wave drag due to lift and vortex drag due to lift. Since in the far-field approach the flow is considered as inviscid, the method is limited by the absence of a method for including the effects of skin friction and leading-edge suction on the aerodynamic force tangential to the wing.

Since both methods require detailed knowledge of the vehicle's geometry– which is usually not available in the conceptual design stage – most methodologies presented in this chapter are derived from linearized theory and classical concepts such as the Sears–Haack equation for slender bodies of revolution, M. Munk's formula for vortex-induced drag due to lift, and R.T. Jones' theory for the pressure drag of slender wings in supersonic flow. Several drag components are estimated from theoretical predictions of minimum drag and refined by correction factors accounting for non-optimum shapes and airframe skin surface roughness. Moreover, several corrections are suggested based on results of non-linear theories that have proved to generate more reliable predictions of lift and drag. Although the present approach does not guarantee an accurate vehicle drag prediction, it can be used to get an insight into the most influential design sensitivities affecting the aerodynamic efficiency of an aircraft configuration. The results are illustrated by examples of shape optimization for isolated wing and fuselage bodies.

7.2 Skin Friction Drag

Some analysts publishing on the design of SCVs have not paid adequate attention to the drag associated with viscosity, perhaps due to the (misplaced) assumption that this subject does not offer new technological insights or challenges. In reality, the skin friction drag of an SCV forms a significant component, typically between 30% and 40% of the total drag. Neglecting this term may result in misleading conclusions, especially in the aerodynamic optimization of vehicle configurations. The observation that much attention is currently paid to the feasibility of natural laminar flow (NLF) to the wing of a supersonic vehicle illustrates that skin friction drag reduction might be a promising technology that should not be neglected.

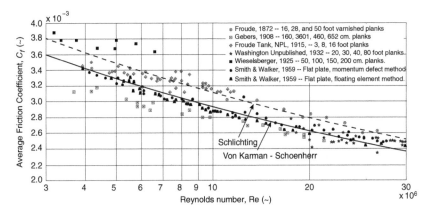

Figure 7.2 Experimental data and interpolation methods on skin friction drag coefficients for turbulent boundary layers along flat plates. Printed courtesy of [5].

7.2.1 Friction Coefficient

The zero-slip condition between the flow and the airplane surface leads to (turbulent boundary layer) shear, caused by viscosity of the airflow. An abundance of theoretical and experimental research on friction drag has been published since the early days of the twentieth century. Figure 7.2 illustrates that considerable scatter was observed between aerodynamic analysis and experimental data, resulting in several approaches for friction drag prediction developed up to the 1970s. An example is the classical Prandtl–Schlichting formula for the friction coefficient due to a turbulent boundary layer along a smooth flat plate at zero angle of incidence to the flow that is applicable to subsonic as well as supersonic flow,

$$C_{\mathrm{F}} = \frac{0.455}{r_{\mathrm{T}}}(\log_{10} \mathrm{Re}_1 - 2.80\log_{10} r_{\mathrm{T}})^{-2.58}, \qquad (7.2)$$

where the Reynolds number refers to the plate length in the flow direction. Denoting the ratio of the adiabatic wall temperature to the static flow temperature, the factor r_{T} accounts for the kinetic heating due to stagnation of the boundary layer as follows:

$$r_{\mathrm{T}} = 1 + P_r^{1/3}\frac{\gamma - 1}{2}M_\infty^2. \qquad (7.3)$$

The Prandtl number P_r defines an index which is proportional to the ratio of energy dissipated by skin friction and the energy transported by thermal conduction. For a turbulent boundary layer in standard conditions $P_r = 0.71$ and hence we may assume $r_{\mathrm{T}} = 1 + 0.178M_\infty^2$. The following alternative for Equation (7.2) is inspired by [13] and [5]:

$$C_{\mathrm{F}} = 0.45(\log_{10}\mathrm{Re}_1)^{-2.58}\left(1 + \frac{\gamma - 1}{2}M_\infty^2\right)^{-0.467}. \qquad (7.4)$$

This equation suggests that the effect of kinetic heating on friction drag at supersonic speed is substantial. For example, the skin friction coefficient decreases by more than 20% during a speed increase from subsonic Mach numbers to Mach 2. And in optimization studies that treat airplane geometry and cruise altitude as design variables, Reynolds number variation may lead to a significant variation of the overall skin friction coefficient of the vehicle.

7.2.2 Flat-plate Analogy

Skin friction drag acts on lifting as well as non-lifting flight vehicle components and is usually predicted in the conceptual design stage by applying the so-called flat-plate analogy. Each component exposed to the flow is represented by a smooth flat plate with same length and area exposed to the flow, situated in undisturbed flow at the same Reynolds number. For instance, Reynolds numbers referred to the fuselage length or the mean geometric chord of a lifting surface are computed for representative flight conditions.

The friction drag area of all vehicle components exposed to the flow is approximated as $(C_D S)_F = K_F C_F S_{wet}$ where the factor K_F accounts for non-ideal drag due to imperfections such as skin surface roughness, seams, control surface slots, and cabin entrance doors. Because an SCV must have an extremely smooth external surface, the surface roughness drag penalty will be smaller than for subsonic airliners. Nonetheless, a higher than average non-ideal drag cannot be avoided for items such as fuselage and tail surfaces. For instance, a typical $K_F = 1.05$ can be assumed for a fuselage and $K_F = 1.15$ for a vertical fin and a horizontal stabilizer. However, wing roughness drag may be neglected, thereby allowing for the short stretch of natural laminar flow behind the leading edge.

The resulting total skin friction drag area is obtained by adding the drag area of all airplane components, $(C_D S)_F = \Sigma K_F C_F S_{wet}$. For instance, the friction drag area of a wing is obtained from multiplication of the friction coefficient according to Equation (7.2) and the net wing area exposed to the flow. It is emphasized that the flow-exposed area of a thin wing is twice the gross area S_w minus the wing area covered by the fuselage.

A less elaborate alternative is to predict the skin friction drag area of a complete airplane configuration is suggested in [4] as follows: $(C_D S)_F = K_F \overline{C}_F \Sigma S_{wet}$. In this approach the mean friction coefficient $\overline{C}_F$ is determined from Equation (7.2) using the Reynolds number referred to the total airplane exposed area ΣS_{wet} divided by the wing span. The factor K_F nominally represents the average drag penalty due to surface imperfections. The estimation of $\overline{C}_F$ leans heavily on the availability of statistical information.

7.2.3 Form Drag

The flow around each flight vehicle component differs from the idealized flat plate flow, which is uniform outside the boundary layer. This deviation is associated with the body volume and consists of a pressure drag and a friction drag increment, together known as form drag. In short, form drag is drag due to adding the boundary-layer displacement thickness to the external physical surfaces. The form drag of a subsonic aircraft wing is substantial due to its relatively thick shape and two-dimensional flow character. However, the lifting surfaces of an SCV are mostly thin and slender, their upper surface friction drag is increased due to the lift distribution but the lower surface friction drag is decreased by a similar amount. Detailed investigations in [2] reveal that the form drag in supersonic flight is of the order of a few percent of the overall skin friction drag. Consequently, the form drag of a slender body in supersonic flow is not considered as an explicit drag component in the present text.

7.3 Slender Body Wave Drag

Selecting the geometry of primary aircraft components is to some extent in the hands of the designer, but the freedom of choice is limited. For example, constraints on the volume required to accommodate the useful load components and their distribution in the aircraft do not allow a great amount of alternatives. The most significant difference between subsonic and supersonic cruising vehicles is the external shape of their major components, which are characterized as slender bodies in the sense that their crosswise dimensions, such as thickness and span, are small compared to their length. In particular, fuselage bodies with a small diameter compared to their length are categorized as slender bodies of revolution, whereas lifting surfaces are slender in the sense that they have a low ratio of span to length.

Any object traveling at supersonic velocity experiences drag caused by shock waves that are mainly generated by the fuselage nose and the leading and trailing edges of the wing and the tailplane. These shock waves are concentrated into conical pressure waves with an expansion in between. This system of waves is propagated without dissipation up to infinite distances from the body, unless they reach the ground where they are reflected and observed in the form of a sonic boom.

7.3.1 Conical Forebody Pressure Drag

The exact solution for the axial flow around the conical fore-body depicted in Figure 7.3 was first published by [7]. In accordance with the oblique flow around a

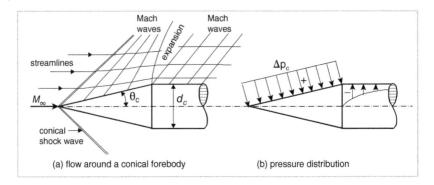

Figure 7.3 Flow around and pressure distribution on a conical fore-body.

two-dimensional wedge described in the Section 4.5, the flow is rotationally symmetrical and the conical flow properties are invariant along rays emanating from the top along the body surface. Different from the (two-dimensional) wedge flow depicted in Figure 4.5, the flow behind the conical shock wave varies away from the cone. In spite of this, the cone experiences a constant pressure increment Δp_c along its surface, which is approximately equal to one third of the overpressure on a two-dimensional wedge for the same values of M_∞ and θ. Written in terms of the pressure coefficient this pressure amounts to

$$(c_p)_c \overset{\text{def}}{=} \frac{p_c - p_\infty}{q_\infty} = \frac{2}{\gamma M_\infty^2}\left(\frac{p_c}{p_\infty} - 1\right). \tag{7.5}$$

The fore-body has a surface area amounting to

$$S_c = \frac{\pi d_c^2}{4\sin\theta_c} \tag{7.6}$$

where d_c and θ_c denote the cone base diameter and its semi-top angle, respectively. The pressure drag increment is determined by

$$\Delta p_c = (\pi d_c^2/4)\sin\theta_c \tag{7.7}$$

and the pressure drag follows from $D_c = \Delta p_c(\pi/4)d_c^2$. Accordingly, the cone drag area amounts to

$$(C_D S)_c = (c_p)_c(\pi/4)d_c^2. \tag{7.8}$$

7.3.2 Von Kármán's Ogive

A body component with low wave drag due to volume was discovered in 1935 by Th. Von Kármán. The "VK ogive" is a fore-body of revolution featuring a pointed nose and a flat base, having minimum pressure drag for specified

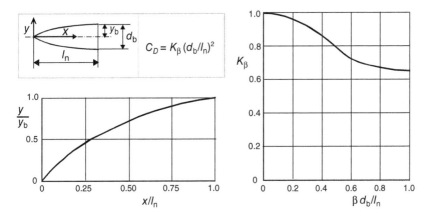

Figure 7.4 Von Kármán's ogive meridian line and variation of the drag coefficient (based on frontal area) with Mach number.

conditions imposed on the nose length l_n, the base area S_b and/or the volume $\mathcal{V}_n$. Figure 7.4 depicts the meridian line of a VK ogive, represented by

$$\left(\frac{y}{y_b}\right)^2 = \frac{2}{\pi}\left[\sin^{-1}\sqrt{X} - (1 - 2X)\sqrt{X(1 - X)}\right] \quad (\text{with} \quad X \stackrel{\text{def}}{=} x/l_n) \quad (7.9)$$

and having $\mathcal{V}_n$ equal to 50% of the volume of its surrounding cylinder. According to linear theory, the pressure drag coefficient at the sonic speed amounts to $C_{D_{wv}} = (d_b/l_n)^2$. Taking into account the variation of the drag with Mach number by means of the factor K_β, the drag area amounts to $C_{D_{wv}}S = K_\beta(16/\pi)(\mathcal{V}_n/l_n^2)^2$. Application of Figure 7.4 appears to offer a practical approach to estimate the pressure drag of a fuselage body, as demonstrated in Section 7.6.

7.3.3 Sear–Haack Body

The wave drag due to volume of slender bodies and wings in supersonic flow is frequently related to the minimum achievable wave drag of slender bodies of revolution. The Sears–Haack (SH) body, discovered independently by W. Haack (1941) and W.R. Sears (1947), represents the unique closed body of revolution with minimum pressure drag at the sonic speed for a specified volume $\mathcal{V}$ and length l over which the volume is spread out longitudinally. The SH body is pointed at both ends and its meridian line (Figure 7.5) is described by

$$\frac{y}{y_m} = \left[4\frac{x}{l}\left(1 - \frac{x}{l}\right)\right]^{3/4}. \quad (7.10)$$

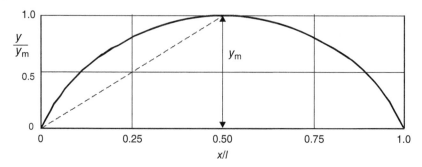

Figure 7.5 Meridian line of the Sears–Haack body.

The volume of an SH body is $3\pi/16$ times the volume of the cylinder surrounding it, and the pressure drag area at Mach 1 is determined by

$$C_{D_{wv}}S = \frac{128}{\pi}\left(\frac{V}{l^2}\right)^2. \tag{7.11}$$

The theoretical minimum drag of the SH body is frequently used as a reference for the wave drag of primary aircraft components. For slender bodies or wings with favorable axial cross sectional area distribution the pressure drag due to volume at low-supersonic speeds is approximated by

$$C_{D_{wv}}S = K_{SH}K_\beta\frac{128}{\pi}\left(\frac{V}{l^2}\right)^2, \tag{7.12}$$

where the factor K_{SH} accounts for the deviation from the ideal SH body shape. Similar to the factor applying to the VK ogive drag (cf. Figure 7.4), the factor K_β represents the variation of the pressure drag with the flow Mach number for slender bodies of revolution. In principle, Equation (7.12) can be used to predict the wave pressure drag of an aircraft wing or a fuselage, but practical SCV shapes are not likely to comply with the SH body shape depicted in Figure 7.5. More attention will be paid to this subject in Sections 7.5 and 7.6.

The flow reversal theorem in linearized potential flow theory introduced in Section 6.2.1 predicts that the pressure distribution on an isolated aft-body is equal to that of an identical fore-body in reverse flow direction. Consequently, an SH body has (nearly) the same pressure drag as two identical VK ogives joined at their base with the same combined volume and length, provided the interference between the flows around the bodies is neglected. This is noteworthy, especially since the meridian lines of the SH body and the joined VK ogives are not identical and, for given total volume and length, the combination of two VK ogives has a larger frontal area than the SH body. More information on the pressure drag of slender bodies of revolution can be found in [17].

7.4 Zero-lift Drag of Flat Delta Wings

The analogy with an SH body can be exploited to show that thin slender wings and even complete vehicle configurations experience a wave drag due to volume that may be substantially less than the wave drag of an SH body with the same length and volume. Their low drag is due to the effect that – different from a body of revolution – the volume of a slender wing is spread out in longitudinal and as well as lateral directions. The resulting $K_{SH} < 1.0$ is obtained on the condition that the derivative of the cross sectional area vanishes at the trailing edge. However, the challenge is to detect a favorable distribution of the longitudinal and lateral cross sectional areas.

Early theoretical methods have been developed to predict the zero-lift wave drag of slender thin wings. Their results can be related to Equation (7.12) and many computational and experimental results have been published to investigate the behavior of K_{SH} for delta wings. Figure 7.6 compares results generated by two theories with a collection of experimental values originating from the development of supersonic theories. In particular, the following empirical equation closely approximates experiments and slender delta wing theory [14]:

$$K_{SH} = 1.17 \frac{1 + 1.5\beta \cot \Lambda_{le}}{1 + 4\beta \cot \Lambda_{le}} \quad \text{for} \quad 0.3 \le \beta \cot \Lambda_{le} \le 1.0, \qquad (7.13)$$

where the leading edge sweep parameter $\beta \cot \Lambda_{le}$ is defined in Figure 6.3.

Theory and wind tunnel data collection of Figure 7.6 suggest that the linear theory is inaccurate for this application and illustrate the necessity to define the wing geometry in more detail than just the leading edge sweep angle. Moreover, Figure 7.6 does not make a distinction between the details of the shape effects represented by K_{SH} and the effect of Mach number variation represented by K_{β}, both featuring in Equation (7.12). More recently, an aerodynamic analysis approach for

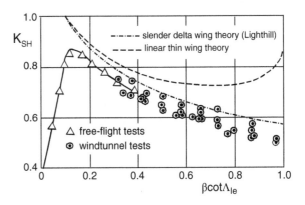

Figure 7.6 Behavior of the factor K_{SH} for delta wings [1].

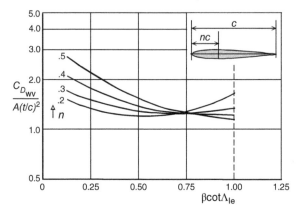

Figure 7.7 Wave drag of delta wings with NACA modified four-digit airfoils [27].

delta wings by means of a comprehensive application of full-potential (non-linear) flow theory in combination with experimental data was published in [27]. Among other things, this publication shows that accurate computation of the wing pressure drag necessitates the introduction of parameters such as the nose shape (sharp or blunt) and the location of the wing profile maximum thickness, illustrated as an example in Figure 7.7. The wave drag coefficient $C_{D_{wv}}$ is presented in terms of the product of the wing aspect ratio A_w and the airfoil thickness ratio squared. The results are generalized by means of the wave drag parameter

$$K_{WV} \overset{\text{def}}{=} \frac{C_{D_{wv}}}{A_w(t/c)^2}. \tag{7.14}$$

The parameter K_{WV} may also be derived from Equation (7.13) by making a few assumptions concerning the wing profile shape. It is then found that the drag factor K_{WV} can be related to the Sears–Haack factor as follows:

$$K_{WV} = 2.2K_{SH} \quad \text{for} \quad \beta \cot \Lambda_{le} \approx 0.75. \tag{7.15}$$

Figure 7.7 depicts one of the results applying to delta wings with NACA modified four-digit series airfoils with rounded leading edge noses, suggesting that for wings with blunt airfoils a typical minimum value for thickness drag is $K_{WV} \approx 1.35$. Although the airfoil maximum thickness location n is an important factor, K_{WV} is nearly independent of n for $0.6 < \beta \cot \Lambda_{le} < 0.8$. This confirms the well known observation that, in order to obtain low thickness wave drag in the cruise condition, the velocity component normal to the leading edge should be close to Mach 0.70.

7.4.1 Drag due to Lift

Analytical means for computing the zero-lift wave drag of wings different from the basic delta shape are not readily available for application in the early stage of

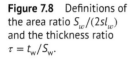

Figure 7.8 Definitions of the area ratio $S_w/(2sl_w)$ and the thickness ratio $\tau = t_w/S_w$.

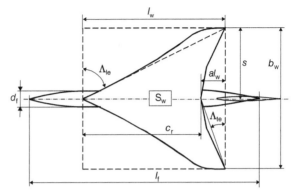

conceptual design. However, Equations (7.12) and (7.14) yield results for delta-like wings with slightly curved or cranked leading edges and/or swept-back trailing edges, which can be useful for an initial design study. For instance, an arrow wing generates less drag than a delta wing with the same volume, planform area, and airfoil shape due to its increased overall streamwise length.

This aspect is elaborated in Section 7.5. Figure 7.8 depicts a wing planform defined by leading and trailing edges extended to the center-line of the fuselage mid-body, which is assumed to have no effect on the total lift. The wing is enclosed by a virtual rectangular box consisting of two identical boxes with sides l_w and semi-span s. The box ratio s/l_w – also known as slenderness ratio – is an essential parameter affecting all contributions to the pressure drag. The area ratio r_S affects the wave drag due to volume and is determined by the shape of the wing edges. For delta and arrow wings with straight leading and straight trailing edges $r_S = 1$; for cranked or curved edges r_S may be (slightly) different from one.

The induced drag of a delta wing with rounded leading edges is obtained from the simple but accurate expression $\Delta C_D/\beta C_L^2 = (\beta C_{L_\alpha})^{-1}$. If more details of the wing geometry are available, the initial estimation of the lift gradient depicted in Figure 5.2 has been refined in Figure 7.4.1. This shows that for wings with blunt leading edges the lift gradient C_{L_α} according to linear theory is in good agreement with experiments for $\beta \cot \Lambda_{le}$ up to 0.5. For higher leading edge sweep angles, the lift gradient – and hence the induced drag – shows a statistical difference up to 10% with experimental data for airfoils with rounded leading edges and up to 25% for sharp airfoils. Although Figure 7.4.1 can be used for an elementary prediction of the induced drag, a more clarifying alternative is to add the vortex drag and the wave drag due to lift according to the following derivations.

7.4.2 Vortex-induced Drag

The minimum vortex-induced drag coefficient $\Delta C_{D_{VL}} = C_L^2/(\pi A_w)$ for a planar wing in subsonic flow is determined by Munk's classical solution [6]. The

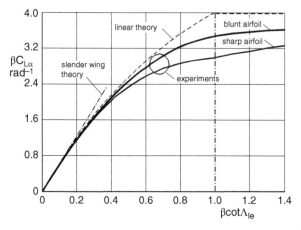

optimality condition corresponds to an elliptical lift distribution along the wing span, resulting in a constant lateral downwash distribution in the Trefftz plane. Jones proved in [11] that the same condition applies to the lower bound of the vortex-induced drag at supersonic speed. According to linearized theory the induced drag of a delta wing amounts to

$$\Delta D_{\mathrm{VL}} = K_{\mathrm{VL}} \frac{(L/b_\mathrm{w})^2}{\pi q} \quad \Rightarrow \quad \Delta C_{D_{\mathrm{VL}}} = K_{\mathrm{VL}} \frac{C_L^2}{\pi A_\mathrm{w}}, \tag{7.16}$$

where K_{VL} accounts for the deviation from the minimum vortex-induced drag. In theory, $K_{\mathrm{VL}} < 1.0$ may be achievable for a non-planar lifting system such as a wing with bent-up or bent-down tips but the practical result for planar wings is a non-elliptic lift distribution with $K_{\mathrm{VL}} > 1.0$. However, assuming $K_{\mathrm{VL}} = 1.0$ is not realistic for wings producing vortex sheets with a conical structure emanating from the leading and/or side edges, as shown in Figure 4.10.

7.4.3 Wave Drag Due to Lift

In 1952, Jones [12] postulated the lower bound for the wave drag due to lift of wings with an elliptical pressure load spread out over the lifting length,

$$\Delta D_{\mathrm{WL}} = \frac{(\beta L/l_\mathrm{w})^2}{2\pi q}, \tag{7.17}$$

suggesting that the streamwise wing length rather than the span is the dominant geometric parameter affecting the wave drag due to lift. Corrected for a non-elliptical longitudinal lift distribution by the factor K_{WL}, the wave drag due to lift is rewritten in terms of its coefficient as follows:

$$\frac{\Delta C_{D_{\mathrm{WL}}}}{\beta C_L^2} = K_{\mathrm{WL}} \frac{\beta A_\mathrm{w}}{8\pi}, \tag{7.18}$$

where $K_{WL} = 1.0$ represents the theoretical lower bound of the wave drag due to lift. It is worth noting that in Equation (7.18) the Mach number appears explicitly in the factor $\beta = \sqrt{M_\infty^2 - 1}$ and hence the wave drag due to lift increases progressively for increasing supersonic Mach numbers. Moreover, Equation (7.18) emphasizes that – different from the vortex-induced drag – the wave drag due to lift increases linearly with the wing aspect ratio.

The theoretical minimum induced drag for planar wings is not obtainable since the lift distribution of a flat delta wing can only approximate the double-elliptic shape, and the factors K_{WL} and K_{VL} are in excess of the theoretical value of 1.0. However, the ideal lift distribution can only be approached for the supersonic design condition by giving a non-planar wing the optimum lateral distribution of camber and twist, resulting in attached flow with full leading-edge suction. Addition of Equations (7.16) and (7.18) yields the induced drag coefficient

$$\frac{\Delta C_D}{\beta C_L^2} = \frac{K_{WL}\beta A_w}{8\pi} + \frac{K_{VL}}{\pi\beta A_w}. \tag{7.19}$$

On the provision that $K_{WL} = K_{VL}$, the induced drag of a delta wing, has a minimum value when the wave drag due to lift and the vortex-induced drag are equal, corresponding to $\beta A_w = 2\sqrt{2}$. This observation forms part of the explanation why the wing of an SCV should have a (considerably) lower aspect ratio than a subsonic transport aircraft wing.

7.5 Wing-alone Glide Ratio

The pressure drag of a wing and fuselage combination may initially be approximated by adding the drag of the two bodies in isolation of each other. The second step is to analyze the interference effects between their surrounding flows and compute the drag reduction or increment associated with this interference. The following text presents an elementary analysis of the glide ratio for isolated delta and arrow wings with straight subsonic leading edges and supersonic trailing edges. Independent variables to be optimized for minimum drag are the aspect ratio and the lift coefficient.

7.5.1 Notched Trailing Edges

Most of the aforementioned information on lift and drag analysis applies to basic delta wings, but a more universal relationship applying to delta as well as arrow wings with straight or slightly curved edges can be found by modifying the geometry of a delta wing as follows:

(a) The center-line section with length c_r remains at a fixed location and the tips are relocated in downstream direction parallel to the center-line, so that the span b_w remains constant. The wing is thus effectively sheared, so that the leading edge sweep angle Λ_{le} is increased and the wing length becomes $l_w = c_r/(1-a)$. Accordingly, the average section thickness ratio t/c, the plan-form area S_w, the aspect ratio A_w, and the volume $\mathcal{V}_w$ remain constant, whereas the box ratio s/l_w decreases by a factor $1-a$.

(b) The area ratio for delta and arrow wings with straight leading and trailing edges amounts to $r_S = 1.0$. For wings with curved or cranked leading and/or trailing edges r_S may be (slightly) different from 1.0.

(c) The wave drag due to volume according to Equations (7.12) or (7.14) and the wave drag due to lift according to Equation (7.16) are reduced by the factor $(1-a)^2$ to account for the overall wing length increment.

Notching the trailing edges has a significant effect on the wave drag due to volume and on the wave drag due to lift. Although increasing the notch ratio appears to be very effective in improving the aerodynamic efficiency, off-design aerodynamics and structural weight constraints impose limitations on the amount of notching, discussed in Chapter 8.

7.5.2 Zero-Lift Drag

The two components of the zero-lift drag are the skin friction drag and the wave drag due to volume. Their matching coefficients for delta and arrow wings with straight leading and trailing edges are obtained from Equations (7.2) and (7.12) or (7.14), corrected for the notch ratio effect, resulting in

$$C_{D_0} = 2K_F C_F + K_{WV}\tau^2 A_w, \quad \text{where} \quad \tau \overset{\text{def}}{=} (t/c)(1-a). \tag{7.20}$$

Equation (7.2) is used to derive the skin friction coefficient C_F of a turbulent boundary layer for an average Reynolds number referred to the mean geometric chord, which may be assumed equal to $l_w(1-a)/2$. The factor K_{WV} can be estimated by means of Section 7.4, in particular Figure 7.7 and Equation (7.14).

7.5.3 Induced Drag

The induced drag ΔD consists of the vortex-induced drag D_{VL} and the wave drag due to lift D_{WL} according to Equations (7.16) and Equation (7.12) or (7.17), respectively. Since the wing span and hence the vortex-induced drag are not affected by

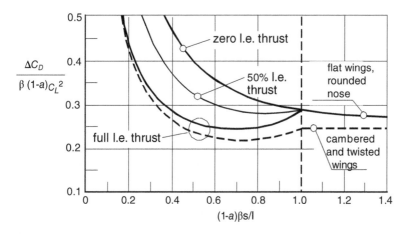

Figure 7.10 Induced drag of straight-tapered delta and arrow wings.

the notch ratio effect, the induced drag coefficient is determined by the following expression:

$$\Delta C_D = \frac{(1-a)\beta C_L^2}{\pi}\left[\frac{K_{WL}(1-a)\beta A_w}{8} + \frac{K_{VL}}{(1-a)\beta A_w}\right]. \tag{7.21}$$

Figure 7.10 suggests that the trailing edge notch effect reduces the induced drag significantly. It is also noteworthy that wing edge sharpness has a significant effect on the attainable leading edge suction and on the induced drag. The ideal condition $K_{VL} = K_{WL} = 1.0$ can only be achieved by carefully designing the camber and twist distribution for the design condition, an essential aspect of aerodynamic design. Consequently, the shape factors K_{WL} and K_{VL} cannot be predicted accurately until detailed information of the wing planform and airfoil geometry, as well as the distribution of camber and warp are available. A realistic assumption for the early design stage is $K_{VL} = K_{WL} \approx 1.15$.

7.5.4 Minimum Glide Ratio

Addition of the drag components according to Equations (7.20) and yields the isolated wing glide ratio, written in terms of the variables A_w and C_L as follows:

$$\frac{C_D}{C_L} = K_F\frac{2C_F}{C_L} + K_{WV}\frac{\tau^2 A_w}{C_L} + K_{WL}\frac{\beta^2(1-a)^2 C_L A_w}{8\pi} + K_{VL}\frac{C_L}{\pi A_w}. \tag{7.22}$$

Although Equation (7.22) should be considered as a provisional result, the notch ratio effects are in accordance with Figure 6.7. Essential parameters affecting the glide ratio are the aspect ratio and the lift coefficient.

- Minimizing the glide ratio with respect to A_w by partial differentiation of Equation (7.22) results in the following solution:

$$(1 - a)\beta A_w = \sqrt{K_{VL}(1 - a)\beta C_L} \left[\frac{K_{WV}\pi\tau^2}{(1 - a)\beta C_L} + \frac{K_{WL}(1 - a)\beta C_L}{8} \right]^{-1/2}.$$

$$(7.23)$$

Equation (7.23) confirms the earlier observation that the glide ratio achieves a minimum value when the leading edges have a sweep angle resulting in a normal flow component $M_\infty \approx 0.7$. Moreover, its application suggests an interesting rule of the thumb that for $\tau^2 \ll 1$ the aspect ratio for minimum drag is obtained from $(1 - a)\beta A_w \approx 2\sqrt{2}$. However, the effect of the wave drag due to volume on the aspect ratio for minimum drag becomes significant for realistic thickness ratios. For instance, the aspect ratio for minimum drag of a Mach 2 aircraft with a thickness ratio of 3% is typically 25% lower than that of a flat plate. This aspect is essential for an integrated wing and body combination, which appears to have a (much) smaller optimum aspect ratio than a combination of discrete wing and body with the same total volume.

- Partial differentiation of Equation (7.22) with respect to C_L yields the lift coefficient for minimum glide ratio

$$(1 - a)\beta C_L = \sqrt{\pi C_{D_0}} \left[\frac{K_{WL}(1 - a)\beta A_w}{8} + \frac{K_{VL}}{(1 - a)\beta A_w} \right]^{-1/2}, \qquad (7.24)$$

where the zero-lift drag coefficient is defined by Equation (7.20).

It is noteworthy that Equation (7.24) yields the classical equation $C_{L_{MD}} = \sqrt{C_{D_0}\pi A_w}$ for the minimum drag of a wing with elliptical lift distribution in subsonic flow for which $K_{VL} = 1$. Equations (7.23) and (7.24) offer closed-form expressions for the partial optima of the aspect ratio and the lift coefficient. An efficient way to obtain the unconstrained minimum glide ratio is an iterative process that normally converges after a few steps due to the fact that the basic terms $(1 - a)\beta A_w$ and $(1 - a)\beta C_L$ appear in both equations. The present section concentrates on the drag of an isolated wing and does not give a solution for a full configuration SCV. Nevertheless, application of Equations (7.22) and (7.23) is useful to observe the following influence of basic geometric parameters on the glide ratio.

1. Different from subsonic transport aircraft featuring high aspect ratio wings, the aspect ratio of a delta wing at supersonic speed has an aerodynamic optimum defined by Equation (7.23), which appears to a have a typical value between 1.5 and 3.0, dependent on the cruise Mach number.

2. The notch ratio forms an essential parameter affecting the drag. For example, a notch ratio of 0.5 for a wing in a flow with Mach 1.6 may bring about an increment of the aerodynamic efficiency up to 40% higher than that of a flat delta wing. There are however essential aerodynamic and structural limitations on the sweep-back angle of the wing trailing edge [29].
3. Equation (7.22) confirms the earlier observation that the skin friction drag contributes a major fraction of the overall drag.

7.6 Fuselage-alone Drag

An SCV has a fuselage volume that is a multiple of the wing volume and its wave drag contributes significantly to the plane's wave drag. However, strict application of a shape similar to the Sears–Haack (SH) body does not result in a practical fuselage configuration in many respects. Figure 7.11 depicts the more realistic schematic shape of a slender body of revolution built up from a pointed fore-body, a cylindrical mid-body (which does not contribute pressure drag), and a pointed aft-body. A circular cylinder is favored for the mid-body because it enables the structure to cope with the cabin overpressure in high-altitude flight and simplifies the arrangement of a flexible passenger cabin layout. In addition, when the aft-body is preceded by a long mid-body, the fore-body does not significantly disturb the flow over the aft-body.

7.6.1 Pressure Drag

According to wave drag analysis based on linearized potential flow theory, the pressure distribution on an aft-body is equal to that of an identical fore-body in

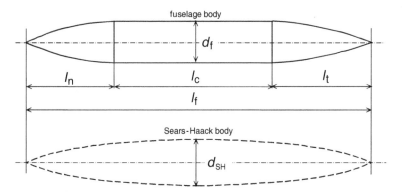

Figure 7.11 Fuselage body with a cylindrical mid-section compared to a Sears–Haack body with the same volume and length.

reverse flow. The fore- and aft-body geometry used for drag prediction of the present slender fuselage concept are VK ogives, introduced in Section (7.3.2). Theoretically, a VK ogive fore-body and an identical aft-body joined at their basis have nearly the same pressure drag as a SH body with the same volume and length, although this concept neglects the interference between the flows around the bodies. A VK ogive has a volume equal to 50% of its surrounding cylinder and hence the fuselage volume is determined from

$$\mathcal{V}_f = \frac{\pi}{4}d_f^2[l_c + 0.5(l_n + l_t)] = K_v\frac{\pi}{4}d_f^2 l_f, \tag{7.25}$$

where K_v denotes the ratio between the fuselage body volume and the volume of its surrounding cylinder. The pressure drag of the combined fore-body and aft-body VK-ogives computed by means of Figure 7.4 amounts to

$$(C_{D_{wv}}S)_f = K_\beta\frac{\pi}{4}d_f^2[(d_f/l_n)^2 + (d_f/l_t)^2], \tag{7.26}$$

where K_β accounts for the (weak) variation of the drag due to non-linearity at speeds in excess of the sonic speed.

An alternative approach for computing the fuselage pressure drag is based on the SH concept for slender bodies with minimum wave drag due to volume introduced in Section 7.3.3. This has a volume $\mathcal{V}_{SH}$, which is a fraction $K_v = 3\pi/16$ of its circumscribed cylinder volume. However, if we require the SH body to have the same volume and length as the fuselage with cylindrical mid-section, its diameter must comply with $d_{SH} = d_f\sqrt{(1 + l_c l_f/24\pi}$. Accordingly, the pressure drag is computed using Equation (7.12), where K_β represents the effect of $M_\infty > 1.0$ for ogives depicted in Figure 7.4, whereas the factor K_{SH} takes into account the fuselage shape deviation from the SH body. Since both drag prediction concepts should yield the same numerical result, comparing Equations (7.12) and (7.26) offers a useful approach to estimate the the factor K_{SH}. This comparison is likely to indicate that a fuselage body with cylindrical mid-section has a considerably higher pressure drag compared with an SH body with the same volume.

7.6.2 Skin Friction Drag

The skin friction drag of a fuselage with cylindrical center section is obtained from its volume $\mathcal{V}_f$ as follows:

$$(S_{wet})_f = 2[\pi\lambda_f(1 + l_c/l_f)\,\mathcal{V}_f^2]^{1/3}, \tag{7.27}$$

whereas the skin friction coefficient C_F for a fully turbulent boundary layer is determined by Equation (7.2) for the Reynolds number based on the fuselage length. After inserting $\mathcal{V}_f$ according to Equation (7.25) and correcting for non-ideal effects due to surface irregularities, the fuselage friction drag area amounts to

$$(C_{D_F}S)_f = 2\,(K_F C_F)_f[\pi\lambda_f(1 + l_c/l_f)\,\mathcal{V}_f^2]^{1/3}. \tag{7.28}$$

7.6.3 Fuselage Slenderness Ratio

The fuselage of an SCV is a typical example of a slender body of revolution. The fuselage slenderness ratio λ_f – defined as the ratio of its overall length l_f to the center-section diameter d_f – is an essential parameter affecting the pressure drag as well as the skin friction drag. Since variation of the slenderness ratio has opposing effects on the drag, optimizing λ_f forms an incentive to reduce the overall body drag.

The fuselage drag is the sum of the pressure drag according to Equation (7.26) and the friction drag according to Equation (7.28). Differentiation with respect to λ_f yields a slenderness ratio for minimum drag $\lambda_f \approx 4K_{SH}^{1/3}K_v^{4/9}(K_FC_F)_f^{-1/3}$. For an SH body with $K_v = 3\pi/16$ and $K_\beta K_{SH} = 1.0$, the slenderness ratio for minimum total body drag is $\lambda_f \approx 3.16(K_FC_F)_f^{-1/3}$, corresponding to a minimum drag area

$$(C_DS)_f = 6.5(K_FC_F)_f^{8/9}V_f^{2/3}. \tag{7.29}$$

The drag of a fuselage with a cylindrical mid-section is considerably higher than that of an SH body with the same volume and length, a penalty that is compensated to some extent by the fact that an SH body requires more volume to carry the same useful load than a fuselage with a mid-body cylinder. Interpretation and application of Equations (7.26) and (7.29) reveal the following trends:

1. The slenderness ratio for minimum drag of an SCV fuselage is at least twice as high as the slenderness ratio of a subsonic airliner fuselage.
2. The skin friction drag and the overall minimum fuselage drag are far less sensitive to variation of λ_f than the pressure drag. For instance, if the slenderness ratio for given volume is varied between 20 and 30, the fuselage drag and the airplane aerodynamic efficiency vary roughly between 1% and 3%. This observation can be helpful in finding an efficient arrangement of the internal cabin layout and passenger accommodation in relation to the diameter and length.
3. The derivation of the minimum fuselage drag is based on numerous assumptions and simplifications. For example, computation of the fuselage pressure drag requires the input of its body volume, the skin friction drag is proportional to the fuselage wetted area and a reliable estimation of the factor K_{SH} requires many details of the geometry.
4. Another reservation is that the derivation of the "optimum slenderness ratio" for minimum drag does not account for the sensitivity of the friction drag coefficient to variation of the Reynolds number.

The present treatment of fuselage drag prediction is valid for an isolated slender body of revolution, whereas the wave drag of a full configuration flight vehicle is sensitive to aerodynamic interference between its main components. Due to the effect of area ruling, the combination of a fuselage with cylindrical midsection and a slender wing may have considerably less drag than the addition of isolated wing and fuselage drag disregarding interference between their surrounding flows. This subject will be treated in Chapter 8.

Bibliography

1 Küchemann, D. *The Aerodynamic Design of Aircraft*. 1st ed. Oxford: Pergamon Press; 1978.

2 Jobe, C.E. Prediction and Verification of Aerodynamic Drag. Part I: Prediction. Thrust and Drag: Its Prediction and Verification. E.E. Covert (ed) *Progress in Astronautics and Aeronautics*. Vol. 98. Reston, VA: Aerospace Research Central; 1985.

3 Jones, R.T. *Wing Theory*. Princeton, NJ: Princeton University Press; 1990.

4 Torenbeek, E. *Advanced Aircraft Design* Wiley and Sons Ltd; 2013.

5 Vos, R., and Farokhi S. *Introduction to Transonic Aerodynamics*. Dordrecht, NL: Springer Science and Business Media; 2015.

6 Munk, M.M. The Minimum Induced Drag of Aerofoils. NACA Report No. 121; 1921. Available at: https://ntrs.nasa.gov/search.jsp?R=19930091456

7 Taylor, G.I., and Maccoll J.W. The air pressure on a cone moving at high speeds. *Proc. R. Aeronautical Soc.* 139(838); 1932.

8 Von Kármán, T. The problems of resistence in compressible fluids. Proc. Fifth Volta Congress, R. Accad. D'Italia; 1936.

9 Jones, R.T. Properties of low-aspect-ratio wings at speeds below and above the speed of sound. NACA Technical Report No. 835; 1946. Available at: https://ntrs.nasa.gov/search.jsp?R=19930091913

10 Sears, W.R. On projectiles of minimum wave drag. *Quarterly Appl. Math.* 4(4):361–366; 1947.

11 Jones, R.T. The minimum drag of thin wings in frictionless flow. *J. Aerospace Sci.* 18(2):75–81; 1951.

12 Jones, R.T. Theoretical determination of the minimum drag of airfoils at supersonic speeds. *J. Aerospace Sci.* 19 (12):813–822; 1952.

13 Sommer, S.C., and Short B.J. Free-flight measurements of turbulent boundary layer skin friction in the presence of severe aerodynamic heating at Mach numbers from 2.8 to 7.0. NACA TN 3391; 1955. Available at: https://ntrs.nasa.gov/search.jsp?R=19930084096

14 Lighthill, M.J. The wave drag at zero lift of slender delta wings and similar configurations. *J. Fluid Mech.* 1(3):337, 1956.

15 Jones, R.T. Aerodynamic Design for Supersonic Speed. *Adv. Aeronautical Sci.* Vol.1. Pergamon Press; 1959.

16 Harris, Jr., R.V. An Analysis and Correlation of Aircraft Wave Drag. NASA TM-X-947;1964.

17 Das, A. Ueber die Berechnung der optimalen aerodynamischen Form von schlanken Flugkörper bei Ueberschallgeschwindigkeit. Jahrbuch 1968 der DGLR, p. 261–285; 1968

18 Küchemann, D., and Weber J. An Analysis of Some Performance Aspects of Various Types of Aircraft Designed to Fly over Different Ranges at Different Speeds. *Progr.Aeronautical Sci.* 9:324–456; 1968.

19 Baals, D.D., Warner Robins A. and Harris Jr. Roy V. Aerodynamic Design Integration of Supersonic Aircraft. *J. Aircraft.* 7(5):385–394; 1970.

20 Kulfan, R.M., and Sigalla A. Real Flow Limitations in Supersonic Airplane Design. AIAA Paper No. 78-147, January 1978.

21 Wright, B.R., Bruckman F. and Radovcich N.A. Arrow Wings for Supersonic Cruise Aircraft. AIAA Paper No. 78-151, January 1978.

22 ESDU. Forebody-afterbody interference wave drag of simple pointed or ducted body shapes with short midbodies. Data Item Bodies S.02.03.08. IHS Markit. 1980.

23 Young, A.D., and Paterson J.H. Aircraft Excrescence Drag. AGARDOgraph No. 264, 1981.

24 ESDU, "Introductory Notes on the Drag at Zero Incidence of Bodies at Supersonic Speeds", ESDU Data Item Bodies S.02.03.01 with Amendments A to C; January 1981.

25 Torenbeek, E. On the Conceptual Design of Supersonic Cruising Aircraft with Subsonic Leading Edges. Delft Progr. Rep. 8:67-80; 1983.

26 Wood, R.M., and Miller D.S. Impact of Airfoil Profile on the Supersonic Aerodynamics of Delta Wings. *J. Aircraft.* 23(9):695; 1986.

27 Wood, R.M. Supersonic Aerodynamics of Delta Wings. NASA Technical Paper 2771; 1988.

28 Bushnell D. Supersonic Aircraft Drag Reduction. AIAA Paper No. 90-1596, June;1990.

29 Kulfan, R.M. Application of Favorable Interference to Supersonic Airplane Design. SAE Paper No. 901988, October; 1990.

30 Seebass, R. Supersonic Aerodynamics: Lift and Drag. RTO AVT Course, Published in Paper RTO EN-4, May; 1998.

31 Ohad Gur, Mason W.H. and Schetz J.A. Full-Configuration Drag Estimation. *J. Aircraft.* 47(4), July-August; 2010.

32 Vos, R. and Vaessen F. A New Compressibility Correction Method to Predict Aerodynamic Interference between Lifting Surfaces. Delft University of Technology; 2015.

8

Aerodynamic Efficiency of SCV Configurations

The present chapter discusses general characteristics of typical configurations designed for supersonic cruising flight and derives criteria for optimizing design selection variables that have significant effect on the aerodynamic efficiency, in particular the configuration slenderness ratio, the wing loading, and the cruise altitude. In the framework of configuration drag analysis and optimization, the term full airframe configuration generally includes a fuselage, a wing, a fore-plane, a horizontal tailplane, and a vertical fin, whereas power plant installation drag is treated as a loss of engine efficiency. The following geometrical characteristics defined in Figure 7.8 have an essential effect on the aerodynamic efficiency of an SCV configuration:

1. The full configuration total volume $\overline{\mathcal{V}}$
2. The configuration slenderness ratio $b_w/\bar{l}$
3. The fuselage slenderness ratio l_f/d_f
4. The wing leading edge sweep-back angle Λ_{le}
5. The wing trailing edge sweep-back angle Λ_{te}
6. The notch ratio $a = \cot \Lambda_{le} / \cot \Lambda_{te}$.

In principle, the drag of isolated primary airplane components cannot simply be added to obtain the full-configuration drag because this approach ignores the aerodynamic interference between bodies that are placed in close proximity with each other. However, in spite of their limited accuracy, the drag prediction methods exposed in Chapter 7 are useful to demonstrate the interaction between the vehicle's overall shape in the conceptual design phase and to identify a maximum or optimum value of the aerodynamic efficiency of an SCV in cruising flight.

8.1 Interaction Between Configuration Shape and Drag

Referring to the expressions derived in Chapter 7, the following relationships are observed between the main drag components and aircraft geometry:

Essentials of Supersonic Commercial Aircraft Conceptual Design, First Edition. Egbert Torenbeek.
© 2020 Egbert Torenbeek. Published 2020 by John Wiley & Sons Ltd.

1. A favorable cross sectional area distribution in the longitudinal direction of a body of revolution results in low pressure drag due to volume, with a theoretical minimum value for a Sears–Haack (SH) body. However, a wing body that is highly stretched in the longitudinal and lateral directions may feature a pressure drag that is up to 40% less than that of an SH body with the same volume. In this respect, an all-wing configuration is likely to beat the aerodynamic efficiency of a discrete wing and fuselage combination.

2. The wave drag due to wing volume and lift decrease when the notch ratio is increased, and sweeping back the trailing edges, can be in the interest of significant drag reduction. However, increasing the slenderness of an arrow wing is subject to aerodynamic limits in subsonic flight and constraints may be necessary to avoid an unacceptable weight penalty due to an excessive structural cantilever ratio.

3. The effect of cruise altitude variation is manifest in the variation of the wing loading and the lift coefficient. Increasing the altitude for a given wing loading and configuration geometry reduces the dynamic pressure, resulting in a reduction in the friction drag area, whereas the lift coefficient – and hence the induced drag – are increasing. In particular, the trade-off between wing size and cruise altitude forms an essential aspect of optimizing the full configuration.

The previous considerations demonstrate that designing for low drag leads to a process of balancing conflicting requirements. For instance: different from the case of an aerodynamically optimized fuselage slenderness ratio, a closed-form solution for an aerodynamically optimized overall configuration is generally not within reach. However, it is known from previous experience that the three concepts depicted in Figure 8.1 are examples of SCV configurations that can be developed to have good aerodynamic characteristics.

Configuration A was studied extensively during the 1960s and the extensive description of their aerodynamic design in [1] concludes that linearized aerodynamic analysis usually results in reliable predictions[1]. The discrete fuselage enclosing the payload in combination with a swept-back arrow wing with subsonic leading edges containing fuel features a favorable longitudinal variation of the cross section according to the principles of the supersonic area rule [5]. Values of the aerodynamic efficiency in supersonic cruising flight up to $L/D \approx 10$ have been quoted in several publications.

Configuration B is an integrated and aerodynamically blended wing concept stemming from the observation that low wave drag can be achieved by spreading out the volume in the streamwise as well as the lateral direction, resulting in leading edges that are swept far behind the Mach cone and hence its optimum slenderness ratio is significantly lower than for configuration A. The blended shape

---

1 Examples of non-linear analysis results are given in Section 7.

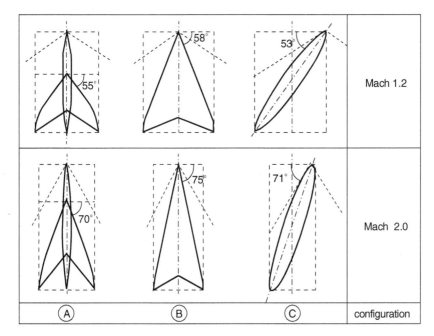

Figure 8.1 Typical configurations for supersonic flight [1].

is also in the interest of reducing the plane's wetted area and the friction drag. Design studies in the framework of the SCAR program during the 1970s and the HSCT program during the 1990s (cf. Chapter 1) suggested that configuration B might achieve $L/D \approx 12$ and can be designed with acceptable operational and commercial characteristics. However, this integrated blended concept is feasible only for a high-capacity transport aircraft.

Configuration C is known as the oblique wing, which transforms a classical non-swept wing into an asymmetric shape with variable sweep dependent on the Mach number. It spreads the volume and lift in the longitudinal as well as the lateral direction by means of a high aspect ratio wing moving oblique to the direction of flight, thereby minimizing vortex-induced drag and wave drag due to lift. Published design studies predict a maximum aerodynamic efficiency between 12 and 15. Reference is made to Chapter 10, which is devoted to the complex set of design aspects of the oblique wing configuration.

8.2 Configuration (A)

Configuration A depicted in Figure 6.8(d) has a discrete fuselage enclosing the payload in combination with a swept-back arrow wing. It is an extension of the typical high-subsonic swept wing aircraft to be used at low-supersonic Mach numbers, whereas varieties of slender delta and arrow wings with curved or cranked leading edges are favored for Mach numbers up to Mach 3. The lowest achievable induced drag occurs when the leading edge is swept well behind the Mach lines and the trailing-edge notch ratio is maximized.

Section 7.5 suggests that the arrow wing has the potential to increase the cruise efficiency of an SCV by at least 10% compared to the configuration with a basic delta wing. Although a straight-tapered arrow wing leads theoretically to the highest possible L/D in supersonic flight, its low-speed properties leave something to be desired and many aspects other than high-speed aerodynamic efficiency have to be considered in the selection process of the best overall planform shape of an arrow wing. Different from supersonic flight, the subsonic L/D degrades when the leading-edge sweep increases, whereas increasing the trailing edge sweep leads to a reduction of the maximum lift coefficient and decay of low-speed performances. The cranked arrow wing concept depicted in Figure 8.1 features un-sweeping outboard wing leading edges in the region of high local up-wash, which improves the subsonic L/D with little detriment to the drag at supersonic speed.

Configuration A has a favorable longitudinal variation of the cross section according to the principles of the supersonic area rule. However, the subsonic linearity of the pitching moment versus the lift curve and the ride quality of an arrow wing improve with increasing leading edge sweep, whereas the opposite applies to increasing the trailing edge sweep. On the other hand, the structural efficiency of a slender arrow wing improves when the trailing edge sweep is decreased, whereas it degrades when the trailing edge sweep is increased. It is therefore unlikely that an analytical optimum design criterion for the trailing-edge sweep exists and the literature suggests that a trailing edge sweep angle higher 30° should be avoided in order to obtain an acceptable aerodynamic design.

The aerodynamic efficiency of configuration A is initially analyzed in its extreme appearance with zero wing thickness and a discrete fuselage containing all useful volume. For this concept it is unavoidable to install lifting surfaces to stabilize and control the airplane. Accordingly, the skin friction drag of the wing is multiplied by a factor K_t, with typical values of 1.10 for a tailless configuration with vertical fin at the rear end of the fuselage or 1.20 for a configuration with a vertical fin and a horizontal tail for stability and control. The resulting drag area at zero lift consists of the wing friction drag area $K_t(C_{D_F})_w S_w$ and the fuselage drag area $(C_D S)_f$, which is obtained from Section 7.6 by adding Equations (7.26) and (7.29). Addition of the four basic drag terms yields the configuration glide ratio:

$$\frac{C_D}{C_L} = \frac{K_t(C_{D_F})_w}{C_L} + \frac{(C_D S)_f}{W/q} + \frac{(1-a)\beta C_L}{2\pi}\left[K_{WL}(1-a)\beta s/l_w + \frac{K_{VL}}{2(1-a)\beta s/l_w}\right].$$

$$(8.1)$$

The most influential variables appearing in Equation (8.1) are $(1 - a)\beta s/l_w$ and $(1 - a)\beta C_L$. The combination of their partial optima defines an unconstrained global optimum of the wing slenderness ratio and the lift coefficient for maximum aerodynamic efficiency, which is very similar to the global optimum of the wing-alone optimum derived in Section 7.5.

8.2.1 Slenderness ratio and lift coefficient for minimum drag

Figure 7.8 illustrates that the slenderness ratio s/l_w – also known as the "box ratio" – is closely related to the wing leading edge sweep angle and the aspect ratio. The aerodynamic slenderness ratio is an essential parameter affecting the aerodynamic efficiency and it is readily shown that for subsonic leading edges $\beta s/l_w$ can vary between zero and one. Hence, the slenderness ratio is selected in the early stage of configuration design.

The variation of the four basic drag components with the slenderness ratio for a specified lift coefficient is depicted in Figure 8.2. According to the initial assumption that configuration A has all its volume in the fuselage, the wing has zero pressure drag and the zero-lift drag coefficient C_{D_0} is not affected by the slenderness ratio. The minimum induced drag for configuration A is obtained from partial differentiation of Equation (8.1) with respect to the slenderness ratio, resulting in

$$\frac{\Delta C_D}{C_L} = \frac{(1 - a)\beta C_L}{2\pi}\sqrt{2K_{WL}K_{VL}} \quad \text{for} \quad (1 - a)\beta s/l_w = \sqrt{\frac{K_{VL}}{2K_{WL}}}. \tag{8.2}$$

For this condition the wave drag due to lift is equal to the vortex-induced drag. Figure 8.2(A) illustrates that in the neighborhood of this condition the glide ratio is not very sensitive to sub-optimum values. An approximation of the minimum glide

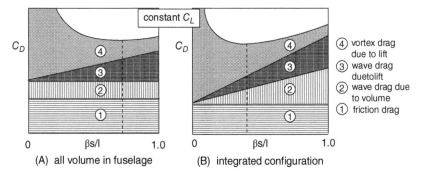

Figure 8.2 Effect of the slenderness ratio on the drag breakdown of different configurations.

ratio is therefore obtained from substitution of Equation (8.2) into Equation (8.1), resulting in

$$\frac{C_D}{C_L} = \frac{C_{D_0}}{C_L} + \left(\frac{\Delta C_D}{C_L}\right)_{\min} = \frac{K_t(C_{D_F}S)_w + (C_D S)_f}{C_L S_w} + \frac{(1-a)\beta C_L}{2\pi}\sqrt{2K_{WL}K_{VL}}.$$

(8.3)

This equation clearly illustrates the significant effect of the notch ratio on drag.

8.2.2 Cruise Altitude for Minimum Drag

The aforementioned results can be applied to explore the influence of varying the wing slenderness ratio and the cruise altitude on the aerodynamic efficiency, for which the lift coefficient is linked with the wing loading and the altitude through the condition of vertical equilibrium. The wing loading is treated as a pre-assigned property, whereas the slenderness ratio and the lift coefficient are considered as the selection variables. Figure 8.3 depicts a fingerprint plot of the aerodynamic efficiency leading to the following observations:

- For $K_{VL} = K_{WL}$, the slenderness ratio for minimum drag (line I) is independent of the lift coefficient: $\beta s/l_w = 1/\sqrt{2}$.

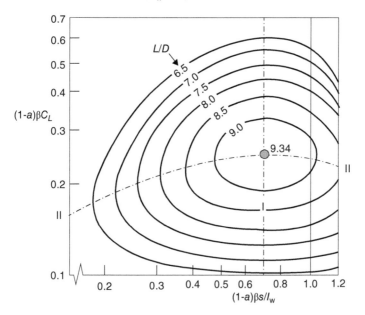

Figure 8.3 Effect of the slenderness ratio and the cruise altitude on the aerodynamic efficiency of wing loading and cruise altitude of an SCV with all volume in the fuselage [9].

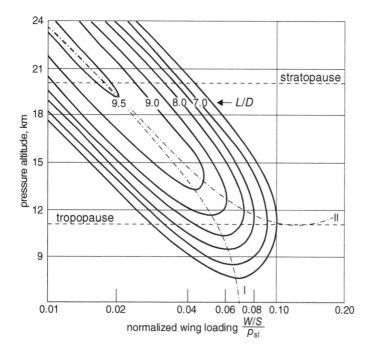

Figure 8.4 Effect of wing loading and cruise altitude on the aerodynamic efficiency of an SCV with all-wing configuration [9].

- The lift coefficient for minimum drag (curve II) is identical to the classical result for subsonic aircraft with a parabolic drag polar: $C_{L_{MD}} = \sqrt{C_{D_0}(\Delta C_D/C_L^2)^{-1}}$.
- The variation of $(L/D)_{max}$ is small – but not insignificant – for $0.5 < \beta s/l_w < 1.0$.

Although the results of this optimization can be useful for an early choice of the cruise altitude, it is emphasized that the altitude and the lift coefficient in cruising flight have an essential influence on the required installed engine thrust, whereas wing size has a considerable effect on its structure weight. Moreover, the cruise altitude has a significant effect on the sonic boom and, hence, cruising at high altitude is a favorable feature for a supersonic cruising aircraft.

8.3 Configuration B

Configuration B shown in Figure 8.1 is an integrated and blended all-wing configuration. Its internal volume for payload and fuel is spread out in longitudinal as well as lateral directions. This supersonic vehicle differs radically from subsonic configurations such as the flying wing concepts promoted by J.K. Northrop during and after World War II and the blended wing-body (BWB), which has been under worldwide study since the 1990s. Different from the extreme case of configuration A with zero-wing volume, the leading edges of configuration B are swept far behind the Mach cone in the interests of the achievable aerodynamic efficiency. Figure 6.8 illustrates that the planform of configuration B can be classified as a delta, a double delta, an arrow, a cranked arrow, or a rhombic wing.

8.3.1 Glide Ratio

Figure 8.2(B) illustrates that the variation of the four basic drag components differs essentially from configuration A. In particular, the substantial wave drag due to volume must be added to the zero-lift drag, making the airframe drag more sensitive to the slenderness ratio. The skin friction drag and the slenderness ratio for minimum drag of configuration B are smaller than for configuration A and the aerodynamic efficiency is more sensitive to sub-optimum deviations. The problem statement for minimizing the drag is summarized by assuming that the wing volume $\bar{V}$ and the planform area $\bar{S}$ are specified quantities following from the airplane top level requirements, whereas the body slenderness ratio (s/l) is considered as the most essential geometric parameter for achieving the maximum aerodynamic efficiency[2]. This input leads to the introduction of a non-dimensional quantity baptized as the equivalent thickness ratio $\bar{\tau} \overset{\text{def}}{=} \bar{V}/(\bar{S}l)$ and to the following modifications of Equation (8.1):

(A) The friction drag of the wing and body combination of configuration A is replaced by the friction drag due to twice the planform area of configuration B and accounting for a vertical fin, resulting in $C_{D_F} = 2K_t K_F C_F$.
(B) The fuselage zero-lift drag of configuration A is replaced by the wave drag due to volume.

The wave drag due to volume is based on the Sears–Haack body pressure drag obtained from Equation (7.12):

$$\frac{C_{D_{wv}}}{C_L} = r_S K_{SH} K_\beta \frac{128\bar{\tau}^2 \beta \bar{s}/\bar{l}}{\pi \beta C_L} \tag{8.4}$$

in which the area ratio r_S has been inserted to allow for curved or cranked leading and/or trailing edges, and Figure 7.6 or Equation (7.13) may be used to estimate the factor K_{SH}. The resulting glide ratio of configuration B is:

$$\frac{C_D}{C_L} = \frac{2K_t K_F C_F}{C_L} + \frac{C_{D_{wv}}}{C_L} + \frac{(1-a)\beta C_L}{2\pi}\left[K_{WL}(1-a)(\beta\bar{s}/l) + \frac{K_{VL}}{2(1-a)(\beta\bar{s}/l)}\right], \tag{8.5}$$

where the term $C_{D_{wv}}/C_L$ equals the second term of Equation (8.4).

---

2 Since all drag contributions relate to the wing, the index w used in the analysis of configuration A is deleted in the present derivation.

Figure 8.2 suggests that the variation of the four basic drag components for configuration B differs considerably from those for configuration A. In particular, its wave drag due to volume is sensitive to the slenderness ratio since the vehicle's length and span depend closely on the geometric slenderness. For the same total vehicle volume, the skin friction drag area of configuration B is smaller and its optimum slenderness ratio is lower than for configuration A, whereas the glide ratio is more sensitive to sub-optimum slenderness ratios. Practical application of Equation (8.5) suggests that configuration B might achieve an aerodynamic efficiency up to 10% higher than configuration A. However, this highly integrated concept may suffer from an unfavorable utilization of the useful internal volume and from high trim drag, whereas in subsonic flight its non-linear aerodynamic characteristics may lead to undesirable flying qualities.

8.3.2 Cruise Altitude and Wing Loading

The drag at moderate supersonic Mach numbers consists of two terms that vary proportional to the dynamic pressure and two terms that vary inversely proportional to the dynamic pressure. The minimum total drag for a given total lift can therefore be expected at an altitude which makes these two couples of forces equal. However, the result of a more detailed analysis is exposed in Figure 8.4, depicting the variation of the aerodynamic efficiency versus the wing loading[3] and the cruise altitude for a blended all-wing configuration with given slenderness ratio. The partial optima indicated in Figure 8.4 for the wing loading (curve I) and the altitude (curve II) approach each other above the stratopause, but the diagram suggests that there exists no absolute unconstrained optimum. For a cruise altitude of 19,000 m the wing loading for the highest $L/D \approx 9.5$ is less than 2% of the wing loading normalized by the sea level standard pressure.

The cruise altitude cannot be optimized by using the aerodynamic efficiency as the essential figure of merit since the installed power plant thrust and the wing size increase sensibly by cruising at increasing altitude. In fact, weight fraction of the fuel plus power plant appears to be a more comprehensive criterion for optimizing the cruise altitude for subsonic transport aircraft as well as supersonic transport aircraft. According to [9] a first approximation of the lift coefficient at the best cruise altitude can be defined in terms of the ratio of power plant weight W_P to fuel weight W_F as follows:

$$\frac{W_P}{W_F} = \frac{1 - (C_L/C_{L_{MD}})^2}{2(C_L/C_{L_{MD}})^3}, \quad \text{where} \quad C_{L_{MD}} = \sqrt{\frac{C_{D_0}}{\Delta C_D/C_L^2}}. \tag{8.6}$$

The improvement in L/D of configuration B seems to be fairly modest compared to configuration A depicted in Figure 8.2. However, configuration A has been analyzed for a wing with zero volume and hence it cannot be considered as a full

---

3 The wing loading in this figure is normalized by dividing it by the sea-level atmospheric pressure.

configuration, the subject of Section 8.4. Nevertheless, it goes without saying that in the advanced design phase an SCV of configuration B will have to be refined considerably relative to the sketch depicted in Figure 8.1. For example, the high-speed all-wing configuration presented in [9] suggests that optimally spreading out the wing volume longitudinally and laterally would result in an average structural height of not more than 70 cm and hence a practical volume distribution would lead to concentration of the volume near the plane of symmetry, making the validity of the drag analysis doubtful.

8.4 Full-configuration Drag

The effect on the glide ratio of varying the slenderness ratio in level flight was considered in Section 8.2 for an SCV with all volume in the fuselage, and in Section 8.3 for a blended all-wing aircraft. A more realistic configuration has its useful volume in both the wing and the fuselage and features lifting surfaces required for stability and control. Derivation of the aerodynamic efficiency of such a concept requires the input of the fuselage volume as well as the wing volume, which are basically derived from the top level requirements and from a prediction of the required fuel weight (Chapter 3). Apart from the empennage drag, the full-configuration drag is based on addition of the wing-alone drag according to Section 7.5 and the fuselage-alone drag according to Section 7.6. However, the analyst must carefully define the areas and volumes exposed to the flow, for instance by making the following assumptions referring to Figure 7.8

(1) The skin friction drag of the wing is determined by the area of the outboard wing sections exposed to the flow and hence the wetted wing area is (approximately) twice the net wing area S_{net}; that is, the gross wing planform area S_w minus the (virtual) wing extension inside the fuselage.

(2) Since the wing wave drag due to volume is determined by the outboard wing sections, the wing section volume inside the fuselage is ignored, and the zero-lift pressure drag computation is based on the aspect ratio of the net wing obtained by putting together the outboard wing sections. For delta and arrow wings with straight leading and trailing edges, the net wing aspect ratio is identical to the aspect ratio of the gross wing, including the (virtual) wing section inside the fuselage body.

(3) The lift generated by the outboard wing sections is carried-over by the fuselage and hence the vortex-induced drag has to be based on the wing span. However, the presence of the fuselage affects the lateral lift distribution, an effect that must be taken into account by correcting the span-efficiency factor K_{VL}. Similarly, the wave drag due to lift must be corrected for the wing–fuselage interference by adapting the numerical value of K_{WL}.

8.4.1 Configuration Glide Ratio

Based on the derivations in Chapter 7 and the assumptions mentioned before, the zero lift drag area of the full configuration is written as follows:

$$C_{D_0}S = [2K_t K_F C_F + K_{WV} r_S(1 - a)\tau^2 A_w](S_w)_{net} + (C_D S)_f, \tag{8.7}$$

where $\tau \overset{\text{def}}{=} (t/c)(1 - a)$ and $K_{WV} \approx 1.35$. Similar to Equation (7.21) the induced drag coefficient amounts to

$$\Delta C_D = \frac{(1 - a)\beta C_L^2}{\pi} \left[\frac{K_{WL}(1 - a)\beta A_w}{8} + \frac{K_{VL}}{(1 - a)\beta A_w} \right], \tag{8.8}$$

where the factors K_{WL} and K_{VL} are corrected for the effect of the fuselage body on the lateral wing lift distribution. Addition of Equations (8.7) and (8.8) leads to the following compact result for the full configuration glide ratio:

$$\frac{C_D}{C_L} = \frac{C_{D_0}}{C_L} + \frac{(1 - a)\beta C_L}{\pi} \left[\frac{K_{WL}(1 - a)\beta A_w}{8} + \frac{K_{VL}}{(1 - a)\beta A_w} \right]. \tag{8.9}$$

Equation (8.9) differs from Equation (7.22) applying to the isolated wing primarily in the addition of the fuselage drag and the drag of lifting surfaces for stability and control, whereas the factor $(S_w)_{net}$ in Equation (8.7) accounts for the reduction of the area exposed to the external flow associated with the assembly of the fuselage and the wing. The full-configuration glide ratio is derived from Equation (8.9), resulting in

$$\frac{C_D}{C_L} = \frac{C_{D_0}}{C_L} + K_{WL} \frac{\beta^2(1 - a)^2 C_L A_w}{8\pi} + K_{VL} \frac{C_L}{\pi A_w} \tag{8.10}$$

and, similar to the optimization treated in Section 7.5, the wing aspect ratio and the lift coefficient are treated as the selection variables.

- Partial differentiation with respect to the wing aspect ratio results in the following condition:

$$(1 - a)\beta A_w = \sqrt{K_{VL}(1 - a)\beta C_L} \left[\frac{K_{WV}\pi r_S \tau^2}{(1 - a)\beta C_L} + \frac{K_{WL}(1 - a)\beta C_L}{8} \right]^{-1/2}. \tag{8.11}$$

Different from the optimum aspect ratio of configuration A treated in Section 8.2.1, the presence of the factor τ^2 has a significant effect on the aspect ratio for minimum drag, which can also be explained by inspection of Figure 8.2.

- The lift coefficient for minimum drag follows from partial differentiation of Equation (8.10), resulting in

$$(1 - a)\beta C_L = \sqrt{\pi C_{D_0}} \left[\frac{K_{WL}(1 - a)\beta A_w}{8} + \frac{K_{VL}}{(1 - a)\beta A_w} \right]^{-1/2}. \tag{8.12}$$

Due to the presence of the wing wave drag due to volume, the zero lift drag coefficient according to Equation (8.9) is considerably higher compared to Equation (8.7) and Equation (7.24) defines a lift coefficient which is considerably higher than $(C_L)_{MD}$ of configuration A depicted in Figure 8.2.2. Since the cruise altitude, the lift coefficient, and the wing loading are interrelated by the condition of vertical equilibrium, the consequences of Equation (8.12) are entirely dependent on the structure of the design problem. In other words: it depends on which variables are considered as (independent) design selection variables and which criterion should be considered as the most significant figure of merit of the SCV.

8.4.2 Notch Ratio Selection

The planform of the wing essentially determines its drag due to lift and wave drag due volume. The leading-edge sweep angle and the notch ratio of configurations A and B are essential design selection variables, as illustrated by Figure 6.7. The induced drag is minimized by sweeping the leading edge well behind the Mach line and maximizing the trailing edge notch ratio. However, similar to the restrictions mentioned for configuration A, many factors other than high-speed aerodynamic efficiency have to be considered in selecting the best overall planform of an arrow wing:

- Different from supersonic Mach numbers, the aerodynamic efficiency in subsonic flow degrades when the leading-edge sweep is increased.
- Increasing the trailing edge sweep leads to a reduction of the maximum lift coefficient and hence to degrading of the low-speed properties.
- The subsonic linearity of the pitching moment versus the angle of incidence and the ride quality of an arrow wing improve with increasing leading-edge sweep. The opposite applies to increasing the notch ratio.
- The structural efficiency of a slender arrow wing improves when the leading edge sweep increases, whereas it tends to degrade with increasing notch ratio.
- Although a pointed arrow wing leads theoretically to the highest possible supersonic L/D, it leaves something to be desired from the local aerodynamic flows [6]. Clipping and unsweeping the wing tips in the region of high local up-wash improve the subsonic aerodynamic efficiency with little detriment to the supersonic induced drag factor.

It is suggested that a trailing-edge sweep angle not more than thirty degrees is selected for an initial planform shape, corresponding to a maximum notch ratio $a = 0.6 \cot \Lambda_{le}$.

8.5   Selection of the General Arrangement

Many SCV configuration design studies have been based on a tailless configuration with relatively low wing loading and aspect ratio – the basic configuration of the Concorde. However, Concorde's general arrangement was based on the condition that the plane should cruise at Mach 2.0, requiring an all-flying slender wing with rather low aspect ratio and featuring elevators for longitudinal control. Figure 8.5 depicts three alternative configurations with different positions of the lifting surfaces for stability and control, referring to the general arrangement of the Mach 1.60 SCT design described in [14]. The following considerations illustrate that selecting the best configuration in the preliminary design stage is a complex process.

8.5.1   Fore-plane Versus After-tail

Figure 8.5 depicts three configurations where the following basic aspects were considered to make a final choice.

Tailless configuration. In supersonic flight a fore-plane operates in undisturbed flow and brings the center of pressure more forward compared to an

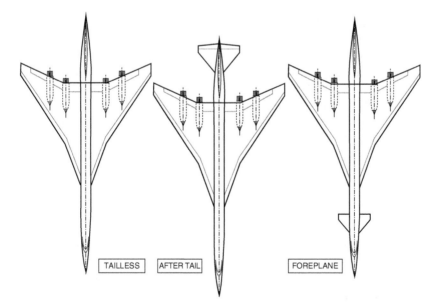

Figure 8.5   Geometry of three configurations designed to cruise at Mach 1.60 [14].

after-tail, thereby reducing the shift of the aerodynamic center associated with increasing the speed from subsonic to supersonic speed. Moreover, a geared fore-plane has a longer moment arm than an after-tail configuration and will be considerably smaller for the same control power, apparently making a fore-plane more attractive than an after-tail configuration.

After-tail configuration

In subsonic flight the induced drag reduction of a high-aspect-ratio wing allows an increased take-off lift coefficient and/or a reduced take-off angle of attack. The lift increment is generated by means of downward deflected trailing edge flaps and – for an inherently stable configuration – the application of a trimming surface is essential to balance the nose-down pitching moment. Hence, the application of a conventional horizontal after-tail with elevator or an all-flying fore-plane becomes attractive. On the other hand, in low speed flight with one inoperative engine the fore-plane configuration has a higher trimmed L/D than the after-tail configuration.

Fore-plane configuration The fore-plane creates down wash reducing the wing's angle of attack. On the other hand the trailing edge flaps increase the wing lift for a given angle of attack. The combined effect is a reduction of the airframe incidence in take-off and landing, offering the possibility to avoid the application of a variable fuselage nose droop angle. This saves a considerable amount of structural weight and maintenance, an essential simplification of flight control systems and a reduction of operational complexity. Different from an after-tail, a fore-plane is destabilizing. Hence, for an inherently stable airframe the conventional tail-plane will permit the wing to be positioned more forward compared to the fore-plane configuration, which is favorable for reducing the fuselage bending and the center of gravity shift.

Different from an after-tail, a fore-plane is destabilizing. Hence, for an inherently stable airframe the conventional tail-plane will permit the wing to be positioned more forward compared to the fore plane configuration, which is favorable for reducing the fuselage bending and the center of gravity shift.

8.5.2 Application of the Area Rule

The component build-up method exposed in the previous section forms a reasonable first approximation of the wave drag due to volume but is inadequate for studying details of the general arrangement's geometry. In particular, the interference between flow fields around primary components such as the fuselage, the wing, tail-plane and engine nacelles has significant effects on the zero-lift drag. A suitable conceptual design method to allow for interference effects is the supersonic area rule method based on slender-body theory [3]. According to this theory the wave drag due to volume of a slender pointed body with length l is determined by

$$C_{D_{wv}} S = \frac{1}{2\pi} \int_0^{2\pi} (C_{D_{wv}} S)_\theta \, d\theta, \tag{8.13}$$

where

$$(C_{D_{wv}} S)_\theta = -\frac{1}{2\pi} \int_0^l \int_0^l S''(x_1, \theta) S''(x_2, \theta) \ln |x_1 - x_2| \, dx_1 \, dx_2. \tag{8.14}$$

Mach planes inclined to the body axis by the Mach angle $\mu = sin^{-1}(1/M_\infty)$ are intersected with the body at variable positions x_1 and x_2. The area S denotes the forward projection of an intersection and the S'' terms are second derivatives of it with respect to the axial position.

The original area ruling method introduced by [3] and explained in detail in [1] is computationally elaborate, but the method proposed by Jumper in [10] is considered as a reasonably accurate alternative, which essentially follows Lomax's wave drag computation [4] but uses the area distribution of the complete configuration as an equivalent body of revolution. The equivalent body drag is then obtained from Mach plane intersections, which are rotationally symmetric. Consequently, for a given axial position and Mach number only one inter-sectional area needs to be computed. Figure 8.6 demonstrates that the wing and tail longitudinal position for the three configurations depicted on Figure 8.5 have an essential effect on the zero lift drag of a full-configuration design. Although the after-tail configuration

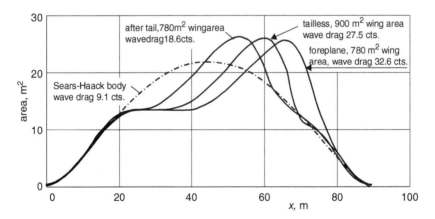

Figure 8.6 Area distribution and wave drag for three configurations of a Mach 1.60 SCV configuration.

has has twice the pressure drag of an SH body with the same total volume, its layout is considerably better than the fore-plane layout with 32.6 counts pressure drag, which is mostly due to the steep cross-sectional gradient aft of the point where the combined fuselage and wing area has its maximum value. And with a zero lift wave drag of 27.5 counts, the tailless configuration has a pressure drag midway between the fore- and after-plane configurations.

In all three shapes there is scope for drag reduction by area ruling the full configuration, which can be effected by negotiating the difference between the SH body shape and the initial configuration area distribution, resulting in a waist fuselage shape. For the after-tail configuration this effect is a cross sectional area increment between approximately 20 m and 40 m and an area reduction between 40 m and 60 m from the fuselage nose.

Bibliography

1 Küchemann, D. *The Aerodynamic Design of Aircraft*. 1st ed. Oxford: Pergamon Press; 1978.

2 Jones, R.T. *Wing Theory*. Princeton NJ: Princeton University Press; 1990.

3 Whitcomb, R. A Study of the Zero-Lift Drag Rise Characteristics of Wing-Body Combinations near the Speed of Sound. NACA RM-L-52H08; 1952. Available at: https://ntrs.nasa.gov/search.jsp?R=19930092271

4 Lomax, H. The Wave Drag of Arbitrary Configurations in Linearized Flow as Determined by Areas and Forces in Oblique Planes. NACA RM A55A18; 1955. Available at: https://ntrs.nasa.gov/search.jsp?R=19930088676

5 Harris, Jr., R.V. An Analysis and Correlation of Aircraft Wave Drag NASA TM-X-947; 1963. Available at: https://ntrs.nasa.gov/archive/nasa/casi .ntrs.nasa.gov/19660029117.pdf

6 Baals, D.D., Warner Robins A. and Harris Jr., Roy V. Aerodynamic Design Integration of Supersonic Aircraft. *J. Aircraft*. 7(5):385–394; 1970.

7 Wright, B.R., Bruckman, F. and Radovcich N.A. Arrow Wings for Supersonic Cruise Aircraft. *J. Aircraft*. 15(12):829; 1978.

8 Squire, L.C. Experimental Work on the Aerodynamics of Integrated Slender Wings for Supersonic Flight. *Progr. Aerospace Sci.* 20(1):1–96; 1981.

9 Torenbeek, E. On the Conceptual Design of Supersonic Cruising Aircraft with Subsonic Leading Edges. *Delft Progr. Rep.* 8 (1983). p. 67–80; 1983.

10 Jumper, E.J. Wave drag prediction using a simplified supersonic area rule. *J.Aircraft*. 20 (10):893–895; 1983.

11 Bushnell D. Supersonic Aircraft Drag Reduction AIAA Paper 90-1596, June;1990.

12 Kulfan, R.M. Application of Favorable Aerodynamic Interference to Supersonic Airplane Design. *SAE Trans.* 99(1):2112–2130; 1990.

13 Wood, R.M., Bauer S.X.S. and Flamm J.D. Aerodynamic Design Opportunities for Future Supersonic Aircraft. ICAS Paper 8.7.1, September; 2002.

14 Torenbeek, E., Jesse E. and Laban M. Conceptual Design of a Mach 1.6 Airliner. AIAA Paper 2004-4541, September; 2004.

9

Aerodynamics of Cambered Wings

Chapters 7 and 8 were intended primarily to explore the basic shapes of the main components and full-airplane configurations offering the promise of high aerodynamic efficiency. Essential characteristics affecting the drag appeared to be the longitudinal volume distribution and the slenderness ratio of the fuselage, the wing thickness ratio, the leading edge sweep angle and the notch ratio. However, this enumeration is arbitrary in the sense that the feasibility of a new SCV configuration is not only determined by its properties in the high-speed cruise condition but also in off-design conditions such as taking-off, subsonic climb and descent, and landing.

The present chapter is intended to give an overview of possibilities to improve the aerodynamic shape of thin wings by applying camber in longitudinal and lateral directions. In particular, reference is made to Figure 7.1, where the possibility of realizing a high percentage of the theoretical maximum leading edge suction is mentioned.

The introduction into the analysis of drag in cruising flight in Chapter 7 is used to make an initial estimate of the aerodynamic efficiency of an SCV in supersonic cruising flight. Figure 9.1 depicts the components of airframe drag at supersonic speeds contributions breakdown of drag components, which does not explicitly mention the leading-edge thrust, an essential contribution to the aerodynamic performance of the wing. Leading-edge thrust is a concentrated force acting in the direction of flight tangent to the wing camber surface forward of the airfoil maximum thickness location and is effectively a reduction of the induced drag, which is critically dependent upon the development of leading-edge suction associated with the flow upward from a lower surface stagnation point around the subsonic leading edge. The following analysis firstly predicts the induced drag of a flat delta wing with sharp leading edges where leading-edge suction cannot develop. It will be followed by an estimation of the theoretical maximum and the practically achievable leading-edge thrust of cambered delta wings.

Essentials of Supersonic Commercial Aircraft Conceptual Design, First Edition. Egbert Torenbeek.
© 2020 Egbert Torenbeek. Published 2020 by John Wiley & Sons Ltd.

total drag			
normal pressure distribution			tangential shear force distribution
vortex-induced drag	shock wave drag		skin friction drag
	due to lift	due to volume	drag due to viscosity
drag due to lift		zero-lift drag	

(leading edge suction ←)

Figure 9.1 Basic drag subdivision at supersonic speeds.

9.1 Flat Delta Wing Lift Gradient and Induced Drag

The lift gradient according to the classical linear theory for flat delta wings with subsonic leading edges was treated in Section 6.4.2 and illustrated in Figure 6.6. Figure 9.2 depicts a collection of experimental data suggesting that for $\beta \cot \Lambda_{le} < 0.4$ the linear theory according to Equation (6.7) is in accordance with experiments, but that the lift gradient for higher values of of the leading edge sweep is significantly overrated. The more accurate non-linear theory according to [18] explains this for airfoils with a rounded nose, where the flow around the nose remains attached for (very) small incidences. However, airfoils with a sharp nose feature an even greater lift loss caused by flow separation at any incidence different from zero and the experimental data for delta wings with sonic and supersonic leading edges show a similar trend of lift gradient over-rating by linear theory. Based on these observations it is proposed in [21]

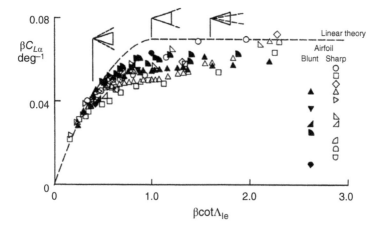

Figure 9.2 Linear theory compared with experiments of thin delta wings [17].

that an acceptable straightforward prediction of the lift gradient is obtained when in Equation (6.7) the elliptic integral $E'(m)$ is replaced by the relationship $1 + k_{nose} m^{1.7}$. Accordingly, the following relationship can be used for the lift gradient:

$$\beta C_{L_\alpha} \approx \frac{2\pi m}{1 + k_{nose} m^{1.7}} \quad \text{for} \quad \beta \cot \Lambda_{le} \geq 1, \tag{9.1}$$

where $k_{nose} = 1.1$ for airfoils with a sharp nose and $k_{nose} = 0.8$ for airfoils with a rounded nose (see Figure 9.3). Alternatively, in terms of the aspect ratio instead of the leading edge sweep – noting that $\beta \cot \Lambda_{le} = \beta A/4$ – the lift gradient for delta wings with subsonic and well-rounded leading edges can be approximated as follows:

$$\beta C_{L_\alpha} = \frac{\pi \beta A/2}{1 + 0.011(\pi \beta A)^{1.7}}. \tag{9.2}$$

According to linear theory, a flat lifting delta wing with subsonic leading edges experiences a flow singularity at the leading edge and the normal force coefficient C_N has the following components in the aerodynamic axis system: $C_L = C_N \cos \alpha$ and $C_D = C_N \sin \alpha$. In reality, in passing from the stagnation point at the lower side to the upper side, the flow will separate at the leading edge and leading-edge suction will not occur. For (very) small incidences, the induced drag is derived from Equation (6.7):

$$\Delta C_D = C_L \sin \alpha \approx \frac{C_L^2}{C_{L_\alpha}} \quad \rightarrow \quad \frac{\Delta C_D}{\beta C_L^2} = \frac{E'(m)}{2\pi m}. \tag{9.3}$$

The magnitude of the leading-edge thrust developed by a thin lifting wing is primarily dependent on the up-wash just ahead of the leading edge and on the airfoil camber just behind the leading edge. If the (thin) wing airfoil has a rounded nose the flow will be able to follow the nose contour and does not

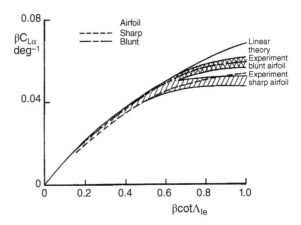

Figure 9.3 Lift gradient of flat delta wings predicted by non-linear theory.

separate under favorable conditions. The suction thus developed at the nose has a forward component in the flow direction, effectively reducing the drag. This can be expressed in terms of a leading-edge thrust coefficient which – according to linear theory – has the following maximum value for fully attached flow:

$$\frac{C_T}{\beta C_L^2} = \frac{\sqrt{1-m^2}}{4\pi m}. \tag{9.4}$$

The induced drag of flat delta wings with full leading edge thrust is thus obtained by subtracting Equation (9.4) from Equation (9.3), resulting in

$$\frac{\Delta C_D}{\beta C_L^2} = \frac{E'(m) - 0.5\sqrt{1-m^2}}{2\pi m}. \tag{9.5}$$

Figures 9.2 and 9.1 show that linear theory over-predicts the lift gradient and hence the induced drag is significantly underrated for typical values of $m > 0.4$ and – to be consistent with the previously proposed lift model – Equation (9.1) is used to predict the induced drag for zero leading-edge thrust as follows:

$$\frac{\Delta C_D}{\beta C_L^2} = \frac{1 + k_{\text{nose}} m^{1.7}}{2\pi m}. \tag{9.6}$$

Assuming the maximum leading-edge thrust according to linear theory to be sufficiently accurate, the minimum induced drag of flat delta wings with blunt airfoils amounts to

$$\frac{\Delta C_D}{\beta C_L^2} = \frac{1 + k_{\text{nose}} m^{1.7} - 0.5\sqrt{1-m^2}}{2\pi m}. \tag{9.7}$$

The proposed Equations (9.6) and (9.7) are compared with the linearized solution in Figure 9.4. The maximum difference between the induced drag of flat and cambered delta wings occurring at a typical subsonic leading edge flow parameter is about is about 10%. This result appears to be confirmed by figure 6.65 of [1], depicting the overall lift dependent factor of flat and cambered slender slender delta wings according to linear theory, as well as experiments showing a difference of more than 30% between cambered delta wings and flat delta wings with zero suction. Accordingly, a realistic estimation of the leading edge thrust is an essential element of drag prediction, which is only feasible by using non-linear analysis.

Figure 9.4 illustrates the relative importance of leading-edge thrust for delta wings with various sweep angles operating at subsonic Mach numbers up to the sonic speed and supersonic Mach numbers up to Mach 2.5. The inset defines the angle of attack α and components of the aerodynamic pressure force acting on a two-dimensional flat wing. The component of the thrust opposing the drag $C_T \cos \alpha$ is depicted as a fraction of the normal force $C_N \sin \alpha$.

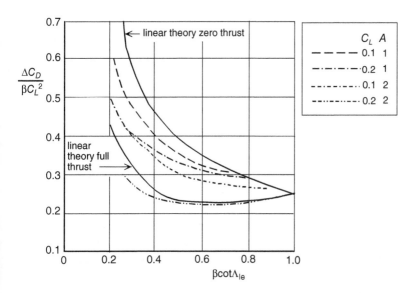

Figure 9.4 Delta wing induced drag predicted by linear and non-linear theory [18].

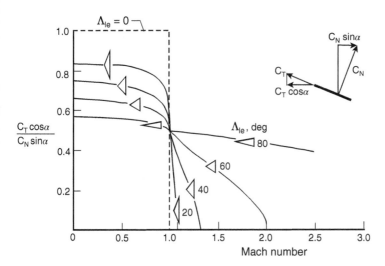

Figure 9.5 Achievable leading-edge thrust for flat delta wings and increased leading-edge thrust by means of [18].

9.1.1 Achievable Leading-edge Thrust

At subsonic speeds the lower sweep angles provide the highest thrust, but at supersonic speeds the more highly swept wings provide the greater relative thrust levels (see Figure 9.4). With increasing subsonic Mach number the fraction $C_T \cos \alpha / C_N \sin \alpha$ approaches one as the sweep angle decreases and the aspect ratio increases, indicating that the leading edge suction force on the airfoil nose negotiates the pressure force on the airfoil tail section. At Mach 1.0 the theory indicates that the component of thrust-opposing drag is one half the drag component of the normal force for all leading-edge sweep angles. With increasing flow Mach numbers there is a steady reduction in the (theoretical) thrust until the leading edge becomes supersonic; that is, $\beta \cot \Lambda_{le} = 1$. The magnitude of the leading-edge thrust developed by a thin lifting airfoil with subsonic leading edges is dependent on the up-wash just ahead of leading edges and on the wing camber just behind the leading edges. The linearized theory predicts that the up-wash and the camber are infinite at the leading edge unless the the wing has a camber surface designed to avoid such a singularity. The influence of up-wash and camber effects is measured by the singularity parameter $\Delta c_p \sqrt{x'}$, where Δc_p and x' denote the lifting pressure coefficient and the distance behind the leading edge where the lifting pressure is acting, respectively. The central problem of determining the maximum leading-edge thrust is the evaluation of the limiting value of the singularity parameter. In-depth studies of this problem and methodologies to estimate the achievable leading-edge thrust can be found in [14] and [20].

9.2 Warped Wings

Publications on the analysis of the leading-edge thrust for delta-like wings can be found in [18] and [20]. Unlike pure delta's, practical wings such as cranked arrow and ogival wings have curved and/or kinked leading and trailing edges, and clipped tips. In high-speed flight conditions their leading edges may be partly subsonic and partly supersonic. If the wing planform to be analyzed is not too dissimilar from the delta shape or the cranked arrow wing shape, it can be replaced by an equivalent arrow wing with a planform shape that can be analyzed with the methods described in Chapters 7 and 8. It can be shown that an elementary relationship can be used to approximate an arbitrary wing shape by a straight-tapered arrow wing plan-form as follows:

- For wings with leading edges consisting of straight sectors the average leading-edge sweep angle is determined by $\cos \Lambda_{le} = 2\Sigma \cos \Lambda_0 c \Delta y / S$, where $c \Delta y$ denotes the plan-form area behind each sector with mean chord c.

- The equivalent root chord amounts to $(c_r)_{eq} = S/b + b/4(1/\cos\Lambda_{le}) - 1/\cos\Lambda_{te})$.
- The equivalent tip chord amounts to $(c_t)_{eq} = S/b - b/4(1/\cos\Lambda_{le})1/\cos\Lambda_{te})$.

The wing leading-edge and the section shape variation have an essential effect on the obtainable leading-edge thrust. In particular, the drag obtained for optimum camber and twist forms an absolute minimum, which is not obtainable in practice since the full theoretical leading-edge thrust cannot be realized and the lift distribution can only approximate the double-elliptic shape. Accordingly, avoidance of flow separation and a high percentage of leading edge thrust can only be obtained for a small range of incidences. The aerodynamic design of a wing depends to a large degree on the SCV's full configuration. In this respect, the difference between the three configurations depicted on Figure 8.1 is essential since the location of the aerodynamic center, the center of pressure and the center of gravity in the design (cruise) condition have to be matched carefully to minimize the plane's trim drag. This subject, discussed in Section 8.5, makes it clear that selecting the best configuration in the preliminary design stage is a complex process. For instance, in the case of Concorde's design the all-wing configuration required a warped wing shape illustrated in Figure 9.6, in which cross-sections and chord-wise sections – including a straight trailing edge – are combined to give a three-dimensional impression of the mean surface at its attachment angle of incidence. The cross sections show pronounced droop of the leading edges required to obtain attached flow in the cruise condition. This geometry results in a mean section with a much larger angle of incidence in the center than at the front than at the rear, giving the required shift of the center of pressure forward from the aerodynamic center.

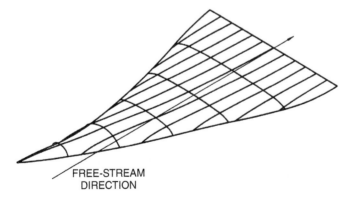

FREE-STREAM
DIRECTION

Figure 9.6 Mean surface of a cambered slender wing [1].

Bibliography

1 Küchemann, D. *The Aerodynamic Design of Aircraft*. Oxford: Pergamon Press; 1978.

2 Anderson, Jr., J.D. *Fundamentals of Aerodynamics*. Int. Ed. New York: McGraw-Hill; 1991.

3 Jones, R.T. The Minimum Drag of Wings in Frictionless Flow. *J. Aeronautical Sci.* 18(2):75–81; 1951.

4 Jones, R.T. Theoretical Determination of the Minimum Drag of Airfoils at Supersonic Speeds. *J. Aeronautical Sci.* 19(12):813; 1952.

5 Emington, E. On the Numerical Evaluation of the Drag Integral. ARC R & M No. 341; 1963. Available at: http://naca.central.cranfield.ac.uk/reports/arc/rm/3341.pdf

6 Harris, R.V. Jr. An Analysis and Correlation of Aircraft Wave Drag NASA-TM-X-947; 1963.

7 Bonner, E. Expanding the Role of Potential Theory in Supersonic Aircraft Design. *J. Aircraft.* 8(5):347–353; 1971.

8 Polhamus, E.C. Predictions of Vortex Lift Characteristics by a Leading-Edge Suction Analogy. *J. Aircraft.* 8(4):193; 1971.

9 Polhamus, E.C. Charts for Predicting the Subsonic Vortex-Lift Characteristics of Arrow, Delta, and Diamond Wings. NASA Technical Note D-6243, 1971.

10 Carlson, H.W., and Miller D.S. Numerical Methods for the Design and Analysis of Wings at Supersonic Speeds. NASA TN D-7713, December 1974.

11 ESDU. Wave Drag of Wings at Zero Lift in Inviscid Flow. ESDU Data Item No. 75004, May. IHS Markit; 1975.

12 Roensch, R.L. Aerodynamic Validation of a SCAR Design. Proc. SCAR Conference, NASA CP-001, Part 1. p. 155–168; 1976.

13 Sotomayer, W.A., and Weeks T.M. Aerodynamic Analysis of Supersonic Aircraft with Subsonic Leading Edges. *J. Aircraft.* 15(7):399–406; 1978.

14 Carlson, H.W., and Mack R.J. Estimation of Attainable Leading-Edge Thrust for Supersonic Wings of Arbitrary Planform. NASA Technical Paper 1270, October; 1978.

15 Carlson, H.W., Mack R.J. and Barger R.L. Estimation of Attainable Leading-Edge Thrust for Wings at Subsonic and Supersonic Speeds. NASA Technical Paper 1500, October; 1979.

16 Robins, A.W., Carlson H.W. and Mack R.J. Supersonic Wings With Significant Leading-Edge Thrust at Cruise. NASA Technical Paper 1632, April; 1980.

17 Wood, R.M. Experimental Investigation of Leading-Edge Thrust at Supersonic Speeds. NASA Technical Paper 2204; 1983.

18 Wood, R.M. Supersonic Aerodynamics of Delta Wings. NASA Technical Paper 2771; 1988.

19 Nelson, C.P. Effects of Wing Planform on HSCT Off-Design Aerodynamics. AIAA Paper No. 92-2629-CP; 1992.

20 Carlson, H.W., McElroy M.O., Lessard W.B. and McCullers L.A. Improved Method for Prediction of Attainable Wing Leading-Edge Thrust. NASA Technical Paper 3557, April; 1996.

21 Torenbeek, E., Jesse E. and Laban M. Conceptual Design and Analysis of a Mach 1.6 European Supersonic Commercial Transport Aircraft, NLR-CR-2003-384, 2003.

10

Oblique Wing Aircraft

One of the unspoken assumptions in aircraft design is that of bilateral or mirror symmetry. For slow flying vehicles this assumption appears to be fully justified. However, once the flight speed exceeds the velocity of sound, the laws of aerodynamics change in such a way as to make it seem inadvisable to arrange the components of an airplane side by side.

R.T. Jones [4]

Since the optimum optimum angle of sweep increases with the flight Mach number, a fixed-wing aircraft has a considerable disadvantage. Hence, the swing wing concept features a wing with two panels having a variable sweep-back angle, which increases with increasing Mach number. Several military supersonic aircraft with swing wings been operational and one of Boeing's early SST projects of the 1960s featured a swing wing (cf. Section 1.2).

An oblique wing – also known as a skewed or slewed wing – was originally proposed by E. de Marcay and E. Moonen in 1912, following the idea to vary the sweep of a wing for landing in side-slip. It was further studied by R. Vogt in Germany for increasing the wing sweep as the speed of aircraft increases. In 1935 A. Busemann pointed out that an infinite swept wing in a supersonic flow is affected only by the flow component normal to the leading edge. In 1958 R.T. Jones noted that wave drag and vortex-induced drag can be minimized by a variable-sweep oblique wing with an elliptic lift distribution, as explained in Section 7.4.1,and he concluded that a wing of infinite extent could fly supersonically without the penalty of wave drag of the flight speed component so that its wing sweep angle becomes dependent of the flight Mach number.

Essentials of Supersonic Commercial Aircraft Conceptual Design, First Edition. Egbert Torenbeek.
© 2020 Egbert Torenbeek. Published 2020 by John Wiley & Sons Ltd.

Figure 10.1 The Ames Dryden AD-1 experimental aircraft. Courtesy: NASA.

10.1 Advantages of the Oblique Wing

Several conceptual designs of a second generation HSCT have featured an oblique wing as the most efficient concept to optimize the sweep angle for each flight speed. The following arguments, illustrated in Figures 10.1 and 10.2 are mostly used to support this opinion.

(A) The subsonic oblique wing demonstrator AD-1 depicted in Figure 10.1 was intended to investigate the flying properties of an airplane with a pivoting oblique wing. By attaching the wing to the fuselage it can be turned to different angles so that flight at different Mach numbers can be made with high aerodynamic efficiency. For flight over land at speeds slow enough to avoid the sonic boom, the wing has a small sweep angle. For flight over water the sweep angle is increased dependent on the optimum Mach number in cruising flight.

(B) The oblique wing has a continuous structure with a pivot in the axial plane of symmetry to turn the wing over an angle dependent on the flight Mach number.

(C) Figure 10.2 illustrates that an oblique wing and a swept wing with the same area and span will have the same vortex-induced drag, whereas the oblique wing will have a considerably lower wave drag due to lift. However, due to its bilateral asymmetry, this concept raises several complications for controlling the aircraft in flight.

Although the benefits of of a highly swept oblique wing were mainly found at transonic and higher speeds, it was recognized that flow compressibility did not have a major influence on many of the problems arising from asymmetry.

Figure 10.2 Geometric advantage of the oblique wing.

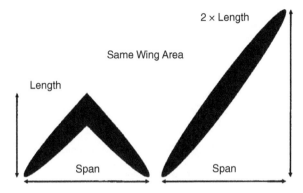

10.2 Practical Advantages of the Oblique Wing

The low-cost, low-speed, low technology AD-1 made its first flight 1979 and successfully demonstrated the concept of a manned aircraft, sweeping the wing to a maximum of 60°. However, the plane experienced cross coupling between the pitching moment and aileron deflection, contributing to unpleasant handling properties at sweep angles above 45°. The oblique wing distributes the lift over twice the wing length compared to conventional, symmetrically swept wings. According to [22] this reduces the wave drag due to lift by a factor of four and the wave drag due to volume by a factor of sixteen.

Varying the sweep angle by turning the wing as a whole has several advantages over the swing wing concept. Turning the oblique wing with a single pivot has considerable practical advantages. Figure 10.3 illustrates the most essential of variable geometry, which demonstrates that the pivot of the obliquely swept continuous wing structure experiences no bending, whereas the swing wing arrangement has wing panels swept back at the wing root where the pivots are subject to high torque and bending loads. Also, sweeping the wing panels backwards for high-speed flight displaces the center of pressure backwards, which compounds the normal rearward center of pressure shift with increasing speed. However, turning the wing of an OWB does not displace the centroid of area and hence the center of pressure relative to the center of gravity.

Figure 10.4(a) illustrates that the structure of the bilaterally symmetric wing with fixed geometry is less favorable because it experiences an unbalanced torsion at the wing root that must be counteracted by a bending moment on the fuselage structure. Moreover, the area rule works advantageously for a straight-beam

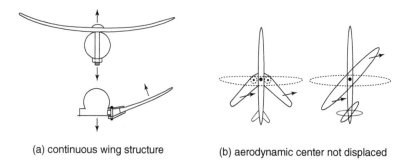

(a) continuous wing structure   (b) aerodynamic center not displaced

Figure 10.3   Advantages of the oblique wing for variable configurations.

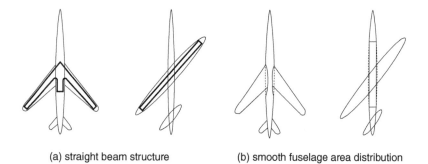

(a) straight beam structure   (b) smooth fuselage area distribution

Figure 10.4   Advantages of the oblique wing for fixed geometry configurations.

oblique wing in combination with a cylindrical fuselage body since a bilaterally symmetric wing requires a highly localized indentation resulting in a larger average fuselage body diameter, an impractical cabin layout and a heavier structure to contain the same number of passengers.

10.3   Oblique Wing Transport Aircraft

An oblique wing and body combination (OWB) consists of a fuselage axially aligned with the direction of flight, a straight wing pivoted to the fuselage, fuselage-mounted engines and lifting surfaces for stability and control. A typical example of a small OWB is depicted in Figure 10.5. Even though the range and fuel burn of the OWB airplane are not much different from that of a conventional configuration, its better low-speed performance due to the wing's reduced yaw angle is attractive. And in addition to its capability to fly efficiently at subsonic speeds it is better suited for small noise-sensitive or terrain-challenged airports due to its short take-off distance and steep climb capabilities. Being able to handle

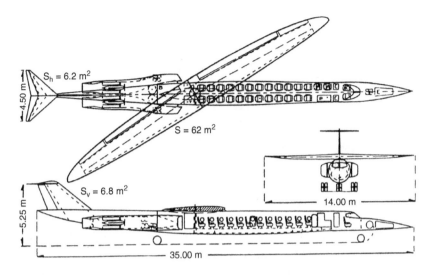

Figure 10.5 Example of a small supersonic oblique wing transport aircraft [12].

air traffic delays, holds and diversions to alternate airports better than a fixed-wing airplane, it should also have more operational flexibility.

The OWB represents one of the simplest implementations of variable geometry and limiting the cruise speed to Mach 1.6 will permit the use of quiet low-bypass turbofans allowing the oblique wing configuration to meet the demands for reducing noise levels around airports as well as the loudness caused by its sonic boom.

10.4 Oblique Flying Wing (OFW)

The unique oblique flying wing (OFW) configuration was first proposed in 1962 by Lee of Handley Page aircraft [2]. More recently, it was pioneered at Stanford University by Jones [9]. Figure 10.6 illustrates an early concept resulting from a Boeing in-house assessment of the OFW. The OFW aircraft consists of a wing containing the payload and the fuel, whereas the engines and vertical tails are hinged to the wing so that their plane of symmetry stays in the flight direction. Dependent on the flight Mach number, the wing's yaw angle is variable during the flight up to (typically) 60° so that the leading edge is swept behind the Mach cone. In this way the plane is enabled to reduce engine noise around airports as well as the sonic boom.

When coupled with an aerodynamically optimal wing thickness ratio of approximately 12%, the wing is inherently able to preserve a straight (cylindrical) wing box, which allows for considerable gain in structural efficiency compared

Figure 10.6 Artist's concept of NASA's oblique flying wing.

to conventional symmetric supersonic wings. To obtain a configuration having minimum drag at supersonic speeds it is necessary to specify the lift and volume within which the external dimensions must be limited. The favorable properties of the OFW requires that in high-speed flight the wing has an angle of yaw such that the Mach number component normal to its long axis is subsonic.

The oblique wing arrangement distributes lift over about twice the length as a conventional swept wing of the same span and sweep, and with an elliptic distribution of the thickness and the chord along the span, its wave drag due to volume of the OFW is equal to that of the symmetrically swept wing having the same lateral span, whereas the vortex-induced drag and the wave drag due to lift are both minimized. This provides a reduction in the wave drag due to volume by a factor of eight. Remarkably, the aerodynamic theory suggests that the straight wing of high aspect ratio which is ideal for low speed flight already has the right shape for the OFW at supersonic speeds.

10.4.1 OFW Flying Qualities and Disadvantages

The primary advantage of the OFW is its low aerodynamic drag at all flying speeds, resulting in excellent aerodynamic performance in subsonic, transonic, and supersonic flight up to $M \approx 1.7$. Due to its much larger length than the conventional arrow-wing airplane the theoretical promise of minimum drag due to the increased longitudinal distribution of lift is very compelling, as are studies which have shown the potential for boom-less flight at Mach numbers up to Mach 1.2. The really significant advantage of the OFW lies in the ease with which the sweep angle can be varied to suit flight conditions. The wing should be non-swept during take-off, landing or holding, and in cruising flight the aerodynamic efficiency may be up to $L/D \approx 20$, which could lead to a very low engine thrust requirement. This configuration will thus minimize the drag in cruising

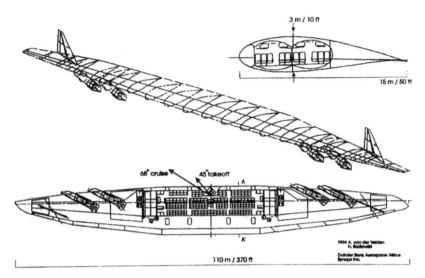

Figure 10.7   Airbus OFW design for a 250 passenger Mach 1.60 transport [19].

flight as well as the unwanted display of energy in the airport environment. While unusual control effects can be prominent in rapid maneuvers, they are not noticeable when controls to maintain steady flight are activated and lack of bilateral symmetry of the HSCT configuration may not be important for cruising flight at high altitude. As a result, the engine integration effort results in lower installed engine cruise thrust and the OFW pays no range penalty for cruising at subsonic speeds. Hence, the range capability of the OFW aircraft depicted in Figure 10.7 may increase by up to 2,000 km if half of the original design mission is flown at Mach 0.95. However, considerable disadvantages of the OFW are the non-ideal shape of the pressure cabin structure required for pressurization and the need for artificial stabilization.

10.5   Conventional and OWB Configurations Compared

Although the OWB configuration does not suffer from the OFW's complications, its aerodynamic efficiency is considerably lower than that of an OFW. The last section of this chapter will therefore be used to compare the aerodynamic efficiency of an oblique wing-body configuration with a second generation HSCT with top level requirements described in Chapter 2. The configuration depicted in Figure 10.8 is designed according to the conventional approach, featuring an arrow wing with a fairly low notch ratio. It is expected to achieve $L/D \approx 10$ in cruising flight – a significantly higher value than Concorde's aerodynamic efficiency.

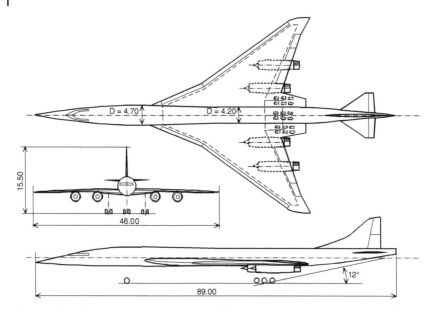

Figure 10.8 Geometry of an arrow-wing HSCT configuration designed to carry 250 passengers over a distance of 10,000 km, cruising at Mach 1.60. [20].

The traditional configuration HSCT will be compared with an OWB configuration designed according to the principles described in Section 10.3. Depicted in Figure 10.9, the design illustrates that the the OWB configuration can do considerably better than just improving the aerodynamic efficiency, based on the following characteristics:

- An essential difference between the conventional and the oblique-wing configurations is the different wing span in the low speed condition. Comparison of Figures 10.8 and 10.9 shows that the OBW wing span is about 40% larger than that of the arrow-wing HSCT. Since the required take-off thrust for a specified take-off field length [3], the AUW and take-off lift coefficient are inversely proportional to the wing span. Hence the OBW needs only three engines instead of four, resulting in a considerably simpler and less expensive installed power plant.
- Both aircraft are assumed to have the same MTOW and cruise altitude. Although the MTOW of the OBW is expected to be considerably less than that of the HSCT, this advantage is not taken into account.

10.5.1 Practical Side-effects

It is often stated that an oblique wing configuration suffers from an aero-elastic problem due to the upward bending of the forward wing section. In reality this

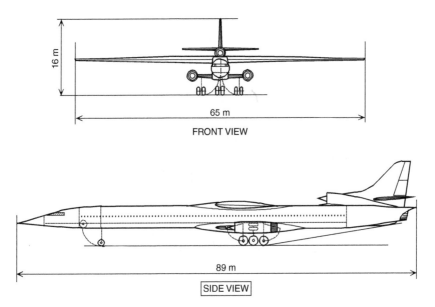

Figure 10.9 Proposed OWB transport aircraft depicted on the airfield.

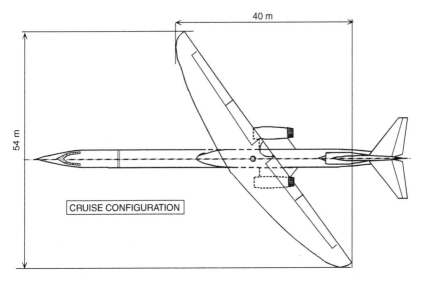

Figure 10.10 The oblique plane configuration in cruising flight.

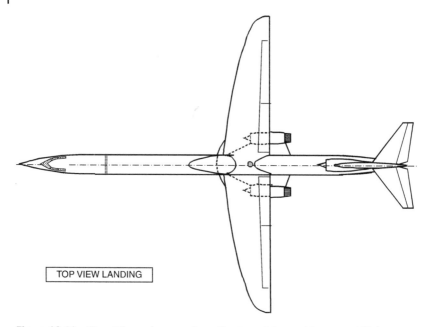

TOP VIEW LANDING

Figure 10.11 The oblique plane configuration in cruising and low-speed flight.

appears to be a minor issue that can be easily coped with by applying a warped wing shape, stiffened structure and/or ailerons on the outer wing half. Moreover, flight tests of the AD-1 have proven that the inherent directional instability of a subsonic oblique wing is easily counteracted, whereas in supersonic flight the effect is expected to be insignificant.

10.6 Conclusion

The drag prediction of the OWB configuration in cruising flight depicted on Figure 10.8 results in the aerodynamic efficiency $C_L/C_D = (0.01/C_L + C_L/8)^{-1}$, corresponding to a maximum efficiency of $(L/D)_{max} \approx 13.5$ for $C_L = 0.25$. This result happens to correspond with the initially assumed cruise altitude of 13,500 m. However, if the plane cruises near the tropopause the aerodynamic efficiency decreases to $L/D \approx 12$, with the result that the cruise drag and the required installed thrust increases by 10%. However, the maximum efficiency in cruising flight is not necessarily the best figure of merit when important other aspects are taken into account such as the effects on the reduced installed thrust and block fuel consumption. The overall optimization of the design is a very complex subject that is outside the contents of this text and it is emphasized that the comparison between the HSCT and the OWB configurations is the first step of a design iteration that does not guarantee that this will be a converging process. Using Equation (3.11) leads to a confirmation that the initially assumed MTOW of 250,000 kg can be realized on the condition that the overall efficiency of the power plant amounts to at least $\eta = 0.40$. Moreover, it is not unlikely that a cruise Mach number up to Mach 1.8 can be achieved with little increase of fuel consumption since a slightly increased lift coefficient has a minor effect on L/D.

The analysis carried out in this chapter proves that designing a second generation HSCT according to transport aircraft having the oblique-wing variable geometry concept offers the opportunity to realize a future supersonic civil transport aircraft with a flight efficiency that is twice as high as that of the Concorde. It is the author's opinion that the oblique wing-body configuration discussed in the present chapter does not suffer from the complexities observed in many failed projects such as those discussed in Chapter 1. Arguably, the OBW transport aircraft cruising at Mach 1.6 has hardly more complex structure and systems than present-day high-subsonic jetliners. The only complication introduced by the oblique wing concept is the installation of the pivoting mechanism and, different from the present high-wing configuration, a low-wing configuration is likely to have a less complicated and lighter pivoting mechanism.

The present analysis shows that the oblique wing configuration has the promise of efficiently cruising at high speed as well as transonic and subsonic speeds, combined with excellent operational capabilities for improving the effects on the environment compared to the presently existing air transport system.

Bibliography

1 Jones, R.T., and Cohen R.T. *High Speed Wing Theory*. Princeton, NJ: Princeton University Press; 1960.

2 G.H. Lee Comments on a paper by D. Kuchemann: Aircraft Shapes and their Aerodynamics. Adv. Aerodynamic Sci. 12:201–218; 1961. Sobieczky, H. (editor), *New Design Concepts for High Speed Air Transport*. Wien: Springer-Verlag; 1997.

3 Torenbeek, E. *Advanced Aircraft Design*. Chichester: John Wiley and Sons Ltd; 2013.

4 Jones, R.T. Reduction of Wave Drag by Antisymmetric Arrangement of Wings and Bodies. *AIAA J.* 10(2):171–176; 1972.

5 Jones, R.T. New Design Goals and a New Shape for the SST. *Astronautics Aeronautics*. December, p. 66–70; 1972.

6 Jones, R.T., and Nisbet J. Transonic Transport Wings – Oblique or Swept? Astronautics Aeronautics. January; 1974.

7 Kulfan, R.M. High-Transonic-Speed Transport Aircraft Study. NASA CR-2465, September; 1974.

8 Nelms, W.P. Applications of Oblique-Wing Technology – An Overview. AIAA Paper 76-943. September; 1976.

9 Jones, R.T. The Oblique Wing – Aircraft Design for Transonic and Low-Supersonic Speeds. *Acta Astronautica.* Vol. 4, p. 99–109. Oxford: Pergamon Press; 1977.

10 McMurty, T.C., Sim A. and Andrews W. AD-1 Oblique Wing Program. AIAA Paper 81-2354, November; 1981.

11 Van der Velden, A.J.M., and Torenbeek E. Design of a Small Supersonic Oblique-Wing Transport Aircraft. *J.Aircraft* 26 (3):193; 1989.

12 Van der Velden, A.J.M. The Aerodynamic Design of the Oblique Flying Wing Supersonic Transport. NASA Contractor Report 177529, May; 1990.

13 Galloway, T., Gelhausen P., Moore M. and Waters M. Oblique Wing Supersonic Transport Concepts. AIAA Paper 92-4230, August; 1992.

14 Waters, M., Ardema M., Roberts C. and Kroo I.M. Structural an Aerodynamic Considerations for an Oblique All Wing Aircraft. AIAA Paper 92-4220, August; 1992.

15 Nelson, C.P. Effects of Wing Planform on HSCT off-design Aerodynamics. AIAA Paper-92-2629, June; 1992

16 Van der Velden, A.J.M. Multi-Disciplinary SCT Design Optimization. AIAA Paper 93-3931, August; 1993 Augustus 1993.

17 Pei Li, Seebass R. and Sobieczky H. The Oblique Flying Wing as the New Large Aircraft. ICAS-96-4.4.2; 1996. Available at: http://www.icas.org/ICAS_ ARCHIVE/ICAS1996/ICAS-96-4.4.2.pdf

18 Rawdon, B.K. Oblique All-Wing Airliner Sizing and Cabin Integration. AIAA Paper 1975-568; 1997

19 Pei Li and Seebass R. Manual Aerodynamic Optimization of an Oblique Wing Supersonic Transport. *J.Aircraft.* 36(6):907; 1999.

20 Torenbeek, E., Jesse E. and Laban M. Conceptual Design and Analysis of a Mach 1.6 Airliner. AIAA Paper 2004-4541; 2004.

21 Wintzer, M., Sturdza P. and Kroo I. Conceptual Design of Conventional and Oblique Wing Configurations for Small Supersonic Aircraft. AIAA Paper 2006-930; 2006.

22 Hirschberg, M., Hart D.M. and Beutner T.J. A Summary of a Half-Century of Oblique Wing Research. AIAA Paper 2007-150; 2007

Index

Essentials of Supersonic Commercial Aircraft Conceptual Design, First Edition. Egbert Torenbeek.
© 2020 Egbert Torenbeek. Published 2020 by John Wiley & Sons Ltd.